Contents

Acknowledgements		iv
Introduction		v
Chapter 1	**Electrical science**	1
	Alternating current theory	2
	Transformers	26
	Instruments and measurement	35
Chapter 2	**Earthing and protection**	43
	Purpose of earthing	44
	Earthing systems	47
	Lightning protection	52
	Electric shock	52
	Earthed equipotential bonding and automatic disconnection (EEBAD)	54
	Residual current devices	54
	The earth-fault loop path	56
	Overcurrent protection	57
	Cable selection	69
Chapter 3	**Lighting**	85
	Incandescent lamps	86
	Regulations concerning lighting circuits	89
	Discharge lighting	90
Chapter 4	**Circuits and systems**	97
	Timers and programmers	98
	Emergency lighting, fire alarms and standby power supplies	100
	Water heating	110
	Space heating	112
	Cooker thermostats and controllers	113
	Closed-circuit television and camera systems (CCTV)	115
	Intruder alarm systems	118
Chapter 5	**Inspection, testing and commissioning**	121
	General requirements	122
	Initial verification, certification and commissioning	127
	The sequence of initial tests	132
	Certificates	152
	Initial inspection	161
	Periodic inspection and testing	171
Chapter 6	**Fault diagnosis and rectification**	187
	Safe working practices for fault diagnosis	188
	General categories of fault	193
	Specific types of fault	200
	Knowledge and understanding of electrical systems and equipment	209
	Stages of diagnosis and rectification	211
	Factors affecting the fault repair process	217
	Special situations	222
Chapter 7	**Motors, generators and motor control**	227
	Concept revision	228
	The d.c. motor	230
	The d.c. and a.c. generator	232
	The a.c. motor	233
	Motor construction	244
	Motor speed and slip calculation	246
	Synchronous a.c. induction motors	249
	Motor windings	250
	Motor starters	251
	Motor speed control	258
Chapter 8	**Electronics**	261
	Resistors	262
	Capacitors	274
	Semiconductor devices	285
	Rectification	293
	Transistors	296
	Integrated circuits (ICs)	305
	Thyristors	306
	Field effect transistors (FETs)	311
Suggested answers		314
Glossary		318
Index		324

Acknowledgements

JTL would like to express its appreciation to all those members of staff who contributed to the development of this book, ensuring that the professional standards expected were delivered and generally overseeing the high quality of the final product. Without their commitment and support – as much to each other as to the project, especially during some of the more fraught days – this project would not have been completed successfully. Particular thanks go to Dave Allan, Brian Tucker, Mike Crossley, Nigel Buckhurst, Phil Cunningham and John Langley. Our thanks also to Paul and Rita Hurt for their patience and assistance during the photoshoot and to John Blaus for his help and advice throughout the project.

The author and publishers are very grateful to the following for assistance in the production of this book and permission to reproduce copyright material:

C.K. Tools (Carl Kammerling International Ltd.)
Paul and Rita Hurt
JTL
Gareth Boden
Ginny Stroud-Lewis

Figure 7.39 on page 256 is reprinted from Electrical Installation Work, 5th edition, Brian Scadden, page 147, Copyright (2005), with permission from Elsevier.

Figure 7.41 on page 257 is reprinted from Advanced Electrical Installation Work, 4th edition, Trevor Linsley, page 97, Copyright (2005), with permission from Elsevier.

The authors and publishers would like to thank the following for permission to reproduce photographs:

AAF-Creatas, p103; Alamy Images, p43, p52, p85; Alamy Images/Daniel Templeton, p1; Alamy Images/Dynamic Graphics /Photis, p261; Art Directors and Trip, p107 (top), p275, p289, p290, p306; Art Directors and Trip/Andrew Lambert, p276; Art Directors and Trip/ H Rogers, p262 (bottom), p292, p296; Corbis/Roger Ressmeyer, p223; Andrew Lambert Photography/Science Photo Library, p233; Ginny Stroud-Lewis, p106, p107 (bottom), p205, p280; Unknown, p97, p121, p150, p151, p187, p227, p234, p236, p241, p244, p256.

All other photos copyright Harcourt Education Ltd/Gareth Boden.

Introduction

What is this book?

This book, and its companion volume, Book 1, have been designed with you in mind. They have a dual purpose:

- To lead you through the City & Guilds 2330 Level 2 and Level 3 Certificates in Electrotechnical Technology (Buildings & Structures)

- To provide a future reference book that you will find useful to dip into, long after you have gained your qualifications.

These books have been specially produced to help you achieve the City & Guilds 2330 at Levels 2 and 3 and between them aim to cover the underpinning knowledge requirement of the relevant NVQ qualification. Both these schemes are available throughout England & Wales and are designed to set a quality standard for learning and training in the electrotechnical sector, whether you are a new entrant to the industry or an experienced worker wishing to update your qualifications.

The books' contents underpin the various topics that you will be examined in as part of the Level 2 and 3 components of the City & Guilds 2330 Technical Certificate, with each chapter concluding with job knowledge tests that will allow you to measure your knowledge and understanding of the various topics.

Qualifications

Generally speaking there are two qualifications that should be considered at this point:

1. *National Vocational Qualifications (NVQs)*: The central feature of any NVQ is the National Occupational Standards (NOS) on which they are based. NOS are statements of performance that describe what competent people in a particular occupation are expected to be able to do … think of it as being a bit like a job description. They cover all the main aspects of an occupation, including current best practice, the ability to adapt to future requirements and the knowledge and understanding that underpins competent performance. Therefore, an NVQ is a qualification that is awarded when individuals can demonstrate that they are actually competent to do the job within the workplace. Consequently, the most valid form of 'evidence of ability' is

by someone watching you do the job at your workplace. However, you will still be expected to demonstrate your knowledge and understanding of these tasks and this is normally measured using external written examinations (such as the Technical Certificate) and oral questioning by an assessor. There are actually five levels of qualification within the NVQ system, with level five being the highest. For the electrical contracting industry, it is currently taken that the NVQ Level 3 is the minimum standard that must be attained for the award of electrician status via the Joint Industry Board (JIB).

2. *Technical Certificate*: For this industry, this is the primary means of providing the underpinning job knowledge and understanding that supports the NVQ competencies. Please note that for apprentices, success is required in both the Level 2 and Level 3 component of the City & Guilds 2330.

As a rough guide, within the electrotechnical industry, meeting the requirements of points 1 and 2 above allow an operative to be able to meet the current industry level for recognition with the JIB as a qualified electrician. However, specific requirements for recognition should always be checked with the JIB.

How this book can help you

There are other key features of this book which are designed to help you make progress and reinforce the learning that has taken place, such features are:

- **Photographs**: easy to follow sequences of key operations.
- **Illustrations**: clear drawings, many in colour, showing essential information about complex components and procedures.
- **Margin notes**: short helpful hints to aid you to good practice.
- **Tables, bullet points and flowcharts**: easy to follow features giving information at a glance.
- **On the job scenarios**: typical things that happen on the job: what would you do?
- **Did you know?**: useful information about things you always wondered about.
- **End of section job knowledge checks**: test yourself to see if you have absorbed all the information, are you ready for the real test?
- **Glossary**: clear definitions and explanations of those strange words and phrases.

Why use this book?

Because it is structured to give you all the basic information required to help you gain the current industry Technical Certificate qualifications and set you on course to an exciting long term career inside a challenging and varied industry … Well done for choosing such a good start!

Chapter 1

Electrical science

OVERVIEW

The work tasks that a competent electrician will undertake are many and varied. However, as well as having practical competence, we also need to know about the operating principles behind the job. The purpose of this chapter is to look at the principles of electrical science and electronics necessary to support electrical and electronic installations. It follows on from the Electrical Science chapter in Book 1 which you may want to refer back to as you work through this chapter. It will cover:

- **Alternating current theory**
- **Transformers**
- **Instruments and measurement**

Alternating current theory

This is one of those subjects that can only be described as theoretical. So you will have to concentrate very carefully and please try to keep awake for this one!

In this section we will be looking at the following areas:

- what is alternating current?
- values of an alternating waveform
- frequency and period
- power factor
- resistance and phasor representation
- inductance
- capacitance
- phasors
- impedance
- resistance and inductance in series (RL)
- resistance and capacitance in series (RC)
- resistance, inductance and capacitance in series (RLC)
- resistance, inductance and capacitance in parallel.
- power in an a.c. circuit
- the power triangle
- three-phase power

What is alternating current?

Alternating current (**a.c.**) is a flow of electrons, which rises to a maximum value in one direction and then falls back to zero before repeating the process in the opposite direction. In other words, the electrons within the conductor do not drift (flow) in one direction, but actually move backwards and forwards.

The journey taken, i.e. starting at zero, flowing in both directions and then returning to zero, is called a cycle. The number of cycles that occur every second is said to be the frequency and this is measured in **hertz** (Hz).

Values of an alternating waveform

If we look at the graph of a sine wave (Figure 1.1), there are several values that can be measured from such an alternating waveform.

Instantaneous value

If we took a reading of induced electromotive force (e.m.f.) from the sine wave at any point in time during its cycle, this would be classed as an instantaneous value.

Average value

Using equally spaced intervals in our cycle (say every 30°) we could take a measurement of current as an instantaneous value. To find the average we would add together all the instantaneous values and then divide by the number of values used. As with the average of anything, the more values used, the greater the accuracy. *For a sine wave only*, we say that the average value is equal to the maximum value multiplied by 0.637. As a formula:

Average current = Maximum (peak) current × 0.637

Peak value

You will remember from Book 1 when the loop in an a.c. generator has rotated for 90° it is cutting the maximum lines of magnetic flux and therefore the greatest value of induced e.m.f. is experienced at this point. This is known as the peak value and both the positive and negative half cycles have a peak value.

Peak to peak value

In Book 1 we said that the maximum value of induced e.m.f., irrespective of direction, is called the maximum or peak value. The voltage measured between the positive and negative peaks is known as the **peak to peak value**. The graph in Figure 1.1 shows this more clearly.

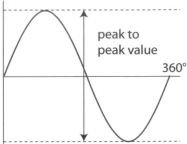

Figure 1.1   Peak to peak value

Root mean square (r.m.s.) or effective value of a waveform (voltage and current)

In Book 1 we showed that in direct current (d.c.) circuits, the power delivered to a resistor is given by the product of the voltage across the element and the current through the element. However, this is only true of the instantaneous power to a resistor in an a.c. circuit.

In most cases the instantaneous power is of little interest, and it is the average power delivered over time that is of most use. In order to have an easy way of measuring power, the r.m.s. method of measuring voltage and current was developed.

The r.m.s. or effective value is defined as being the a.c. value of an equivalent d.c. quantity that would deliver the same average power to the same resistor.

When current flows in a resistor, heat is produced. In Book 1 you saw that, when it is direct current flowing in a resistor, the amount of electrical power converted into heat is expressed by the formulae:

$$P = I^2 \times R \quad \text{or} \quad P = V \times I$$

However, an alternating current having a maximum (peak) value of 1 A does not maintain a constant value (see Figure 1.1). The alternating current will not produce as much heat in the resistance as will a direct current of 1 A. Consider the circuits in Figure 1.2.

Figure 1.2   d.c. and a.c. circuits

In both the circuits in the figure, the various supplies provide a maximum (peak) value of current of 1 A to a known resistor. However, the heat produced by 1 ampere of alternating current is only 70.7°C compared to the 100°C of heat that is produced by 1 ampere of direct current. We can express this using the following formula:

$$\frac{\text{Heating effect of 1 A maximum a.c.}}{\text{Heating effect of 1 A maximum d.c.}} = \frac{70.7}{100} = 0.707$$

Therefore, **the effective or r.m.s. value of an a.c. = $0.707 \times I_{max}$**

where I_{max} = the peak value of the alternating current.

We can also establish the maximum (peak) value from the r.m.s. value with the following formula:

$$I_{max} = I_{r.m.s.} \times 1.414$$

The rate at which heat is produced in a resistor is a convenient way of establishing an effective value of alternating current, and is known as the 'heating effect' method.

An alternating current is said to have an effective value of one ampere when it produces heat in a given resistance at the same rate as one ampere of direct current.

To see where the r.m.s. value sits alongside the peak value, look at the following figures.

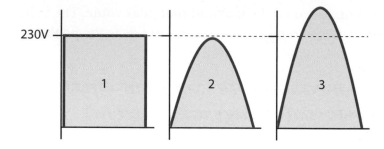

Figure 1.3 Peak value diagrams

The wave in Figure 1.3(1) represents a 230 V d.c. supply, running for a set period of time, with the heating effect produced shown as the shaded area.

The wave in Figure 1.3(2) represents an AC 50 Hz supply in which the voltage peaks at 230V during one half-cycle, with the heating effect produced shown as the shaded area. As the wave only reaches 230V for a small period of time, less heat is produced overall than in the d.c. wave.

The wave in in Figure 1.3(3) represents an increased peak value of voltage to give the same amount of shaded area as in the d.c. example. In this diagram 230 V has become our r.m.s. value.

Important Note: Unless stated otherwise, all values of a.c. voltage and current are given as r.m.s. values.

Frequency and period

Remember from Book 1 that the number of cycles that occur each second is referred to as the frequency of the waveform and this is measured in hertz (Hz). The frequency of the UK supply system is 50 Hz.

In the basic arrangement of the a.c. generator loop, if one cycle of e.m.f. was generated with one complete revolution of the loop over a period of one second, then we would say the frequency was 1 Hz. If we increased the speed of loop rotation so that it was producing five cycles every second, then we would have a frequency of 5 Hz.

We can therefore say that the frequency of the waveform is the same as the speed of the loop's rotation, measured in revolutions per second. We can express this using the following equation:

Frequency (f) = Number of revolutions (n) × Number of pole pairs

If we apply this to the simple a.c. generator and rotate the loop at 50 revolutions per second, then:

Frequency = 50 × 1 (there is 1 × pole pair) = 50 Hz

The amount of time taken for the waveform to complete one full cycle is known as the periodic time (T) or period. Therefore, if 50 cycles are produced in one second, one cycle must be produced in a fiftieth of one second. This relationship is expressed using the following equations:

$$\text{Frequency (f)} = \frac{1}{\text{Periodic time}} = \frac{1}{T} \qquad \text{Periodic time (T)} = \frac{1}{\text{Frequency}} = \frac{1}{f}$$

Power factor

When we are dealing with a.c. circuits, we are often looking at the way power is used with particular types of component within the circuit.

Generally speaking, **power factor** is a number less than 1.0, which is used to represent the relationship between the apparent power of a circuit and the true power of that circuit. In other words:

$$\text{Power Factor (PF)} = \frac{\text{True Power (PT)}}{\text{Apparent Power (PA)}} \qquad \text{Or: } PF = \frac{PT}{PA}$$

In terms of units:

$$\text{Power Factor (PF)} = \frac{\text{Power (W)}}{\text{Voltage} \times \text{Current (VA)}}$$

Power factor has no units, it is a number. It is also determined by the phase angle, which we will cover shortly.

We will be coming back to this quite frequently over the next few pages, so by the end of the section you will have a good understanding of power factor.

Resistance (R) and phasor representation

Whereas the sine wave is useful, it is also difficult and time consuming to draw. We can therefore also represent a.c. by the use of phasors. A **phasor** is a straight line whose length is a scaled representation of the size of the a.c. quantity and whose direction represents the relationship between the voltage and current, this relationship being known as the phase angle.

To see briefly how we use phasors, let us look at the following diagram, where a tungsten filament lamp has been included as the load.

Circuits like this are said to be **resistive**, and in this type of circuit the values of e.m.f. (voltage) and current actually pass through the same instants in time together. In other words, as voltage reaches its maximum value, so does the current (see Figure 1.4).

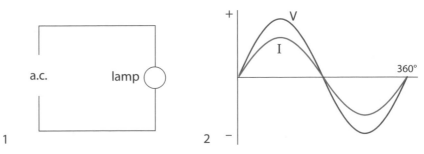

Figure 1.4 Circuit and sine wave diagrams

This happens with all resistive components connected to an a.c. supply and as such the voltage and current are said to be '**in phase**' with each other, or possess a zero phase angle.

The graph in Figure 1.4 shows this when represented by a sine wave. However, we could also show this by using a phasor diagram as shown in Figure 1.5.

Figure 1.5 Phasor diagram for zero phase angle

We can therefore say that a resistive component will consume power and we would carry out calculations as we would for a d.c. circuit (i.e. using $P = V \times I$).

We can also say that resistive equipment (filament lamps, fires, water heaters) use this power to create heat, but such a feature in long cable runs, windings etc. would be seen as unsuitable power loss in the circuit (i.e. using $P = I^2R$).

Inductance (L)

If the load in our circuit were not a filament lamp, but a motor or transformer (something possessing windings), then we say that the load is **inductive**.

The sine wave used to represent this inductive circuit would look like this

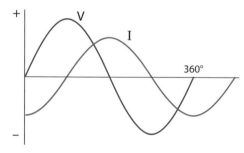

If we represented this as a phasor diagram we end up with

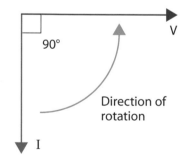

Figure 1.6 Sine wave and phasor diagrams for inductive circuit

With an inductive load the voltage and current become '**out of phase**' with each other. This is because the windings of the equipment set up their own induced e.m.f., which opposes the direction of the applied voltage, and thus forces the flow of electrons (current) to fall behind the force pushing them (voltage). However, over one full cycle, we would see that no power is consumed. When this happens, it is known as possessing a lagging phase angle or power factor.

As voltage and current are no longer perfectly linked, this type of circuit would be given a power factor of less than 1.0 (perfection), for example 0.8.

As we can see from Figure 1.6, the current is lagging the applied voltage by 90°. To make things easier in this exercise, we assumed that the above circuit is purely inductive. However, in reality this is not possible, as every coil is made of wire and that wire will have a resistance. The opposition to current flow in a resistive circuit is resistance.

The limiting effect to the current flow in an inductor is called the **inductive reactance**, which we are able to calculate with the following formula:

$X_L = 2\pi fL \ (\Omega)$

where:

X_L = inductive reactance (ohms – Ω)

f = supply frequency (hertz – Hz)

L = circuit inductance (henrys – H)

Let us now look at inductive and resistive circuits and see if the current is affected by a lagging phase angle.

To recap, we said that power factor is the relationship between voltage and current and that the ideal situation would seem to be the resistive circuit, where both these quantities are perfectly linked.

In the resistive circuit we know that the power in the circuit could only be the result of the voltage and the current ($P = V \times I$). This is known as the apparent power, and possesses what we call unity power factor, to which we give the value one (1.0).

However, we now know that depending upon the equipment, the true power (actual) in the circuit must take into account the phase angle and will often be less than the apparent power … but never greater.

True power (in watts) is calculated using the cosine of the phase angle (cos Ø). The formula is:

$P = VI \cos Ø$ (remember we do not have to use the '×' sign)

When there is no phase lag, Ø = 0 and cos Ø = 1, a purely resistive circuit. To prove our previous points, let us consider the following.

Example

If we have an inductive load, consuming 3 kW of power from a 230 V supply, with a power factor of 0.7 lagging, then the current (amount of electrons flowing) required to supply the load is:

$$P = V \times I \times \cos Ø$$

Cos Ø = Power Factor. Therefore by transposition:

$$I = \frac{P}{V \times \cos Ø}$$

Or in other words:

$$I = \frac{P}{V \times PF}$$

Therefore:

$$I = \frac{3000}{230 \times 0.7}$$

And so:

$$I = 18.6\,A$$

However, if the same size of load was purely resistive, then cos Ø = 0, thus the power factor would be 1.0, and thus:

$$P = V \times I \times PF$$

Therefore by transposition:

$$I = \frac{P}{V \times PF}$$

Therefore:

$$I = \frac{3000}{230 \times 1}$$

And consequently:

$$I = 13\,A$$

In other words, the lower the power factor of a circuit, then the higher the current will need to be to supply the load's power requirement.

It therefore follows that if power factor is low, then it will be necessary to install larger cables, switchgear etc. to be capable of handling the larger currents. There will also be the possibility of higher voltage drop due to the increased current in the supply cables.

Consequently, local electricity suppliers will often impose a financial fine on premises operating with a low power factor. Fortunately, we have a component that can help. It is called the capacitor.

Capacitance (C)

Simply put, a **capacitor** is a component that stores an electric charge if a potential difference is applied across it.

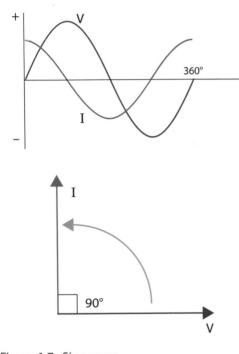

Figure 1.7 Sine wave and phasor diagrams for capacitive circuits

The capacitor's use is then normally based on its ability to return that energy back to the circuit. When a capacitor is connected to an a.c. supply, it is continuously storing the charge and then discharging as the supply moves through its positive and negative cycles. But, as with the inductor, no power is consumed.

This means that in a capacitive circuit, we have a leading phase angle or power factor. The sine wave and phasors used to represent this would look as in Figure 1.7.

As we can see from the previous diagram, the current leads the voltage by 90°. Consequently, the capacitor is able to help because it provides a leading power factor, and therefore if we connect it in parallel across the load, it can help neutralise the effect of a lagging power factor.

The opposition to the flow of a.c. to a capacitor is termed capacitive reactance, which like inductive reactance, is measured in ohms and calculated using the following formula:

$$X_C = \frac{1}{2\pi fC} \ (\Omega)$$

where:

X_C = capacitive reactance (ohms – Ω)

f = supply frequency (hertz – Hz)

C = circuit **capacitance** (farads – F)

Since, in this type of circuit, we have voltage and current but no real power (in watts), the formula of $P = V \times I$ is no longer accurate. Instead, we say that the result of the voltage and current is reactive power, which is measured in reactive volt amperes (VAr).

The current to the capacitor, which does not contain resistance or consume power, is called reactive current.

Well, so far in our attempt to explain power factor, we have looked at a range of different subjects including resistance, inductance, capacitance and also talked about phasor diagrams. As if that wasn't bad enough, some circuits contain combinations of these components.

Sadly, in order to be able to work out these calculations, you will need to use both the trigonometry knowledge you gained in Book 1 and also the following small section regarding the addition of phasors.

Phasors

When sine waves for voltage and current are drawn, the nature of the wave diagram can be based upon any chosen alternating quantity within the circuit. In other words, we can start from zero on the wave diagram with either the voltage or the current.

In electrical science we often need to add together alternating values. If they were 'in phase' with each other, then we would simply add the values together. However, when they are not in phase we cannot do this, hence the need for phasor diagrams.

When we use phasor diagrams the chosen alternating quantity is drawn horizontally and is known as the reference.

When choosing the reference phasor, it makes sense to use a quantity that has the same value at all parts of the circuit. For example in a series circuit the same current flows in each part of the circuit, therefore use current as the reference phasor. In a parallel circuit, the voltage is the same through each branch of the circuit and therefore we use voltage as the reference phasor.

Using the knowledge gained in the previous section, we can now measure all phase angles from this reference phasor.

Our answer, or resultant, is then found by completing a parallelogram. If we use Figure 1.8 as an example, we have been given the values of phasor A and phasor B. Therefore, the result of adding A and B together will be phasor C.

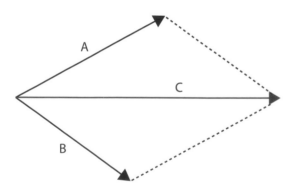

Figure 1.8   Phasor diagram

Impedance

Previously, we have been discussing components with an a.c. circuit. In fact, what those components are actually offering is opposition to the flow of current. By way of a summary, we could say that we now know that:

- The opposition to current in a resistive circuit is called resistance (R) is measured in ohms and the voltage and current are in phase with each other.
- The opposition to current in an inductive circuit is called inductive reactance (X_L), is measured in ohms and the current lags the voltage by 90°.
- The opposition to current in a capacitive circuit is called capacitive reactance (X_C), is measured in ohms and the current leads the voltage by 90°.

However, we also know that circuits will contain a combination of these components. When this happens we say that the total opposition to current is called the **impedance** (Z) of that circuit.

In summary:

- The power consumed by a resistor is dissipated in heat and not returned to the source. This is called the **true power**.
- The energy stored in the magnetic field of an **inductor** or the plates of a capacitor is returned to the source when the current changes direction.
- The power in an a.c. circuit is the sum of true power and reactive power. This is called the **apparent power**.
- **True power is equal to apparent power in a purely resistive circuit** because the voltage and current are in phase. Voltage and current are also in phase in a circuit containing equal values of inductive reactance and capacitive reactance. If the voltage and current are 90 degrees out of phase, as would be the case in a purely capacitive or purely inductive circuit, the average value of true power is equal to zero. There are high positive and negative peak values of power, but when added together the result is zero.
- Apparent power is measured in volt-amps (VA) and has the formula: $P = VI$
- True power is measured in watts and has the formula $P = VI \cos Ø$
- **In a purely resistive circuit** where current and voltage are in phase, there is no angle of displacement between current and voltage. The cosine of a zero degree angle is one, and so, the power factor is one. This means that all the energy that is delivered by the source is consumed by the circuit and dissipated in the form of heat.
- **In a purely reactive circuit**, voltage and current are 90 degrees apart. The cosine of a 90 degree angle is zero so the power factor is zero. This means that the circuit returns all the energy it receives from the source, back to the source.
- **In a circuit where reactance and resistance are equal**, voltage and current are displaced by 45 degrees. The cosine of a 45 degree angle is 0.7071, and so, the power factor is 0.7071. This means that such a circuit uses approximately 70 per cent of the energy supplied by the source and returns approximately 30 per cent back to the source.

Resistance and inductance in series (RL)

Consider the following diagram.

Figure 1.9 Resistor and inductor in series

Here we have a resistor connected in series with an inductor and fed from an a.c. supply. In a series circuit, the current (I) will be common to both the resistor and the inductor, causing **voltage drop** V_R across the resistor and V_L across the inductor.

The sum of these voltages must equal the supply voltage. Here's how to construct a phasor diagram for this circuit.

In a series circuit we know that current will be common to both the resistor and the inductor. It therefore makes sense to use current as our reference phasor. We also know that voltage and current will be in phase for a resistor. Therefore, the volt drop (**p.d.**) V_R across the resistor must be in phase with the current. Also, in an inductive circuit, the current lags the voltage by 90°.

If the current is lagging voltage, then we must be right in saying that voltage is leading the current.

This means in this case that the volt drop across the inductor (V_L) will lead the current by 90°. We can then find the value of the supply voltage (V_S), by completing the parallelogram that was discussed on page nine. When we draw phasors, we always assume that they rotate anti-clockwise and the symbol Ø represents the phase angle.

There are two ways of doing the drawing.

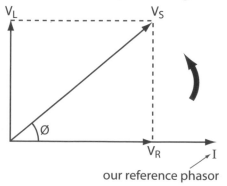

Figure 1.10 Two ways of drawing a phasor diagram

We can see that in the second example, the phasors produce a right-angled triangle. We can therefore use Pythagoras' Theorem to give us the formula:

$$V_S^2 = V_R^2 + V_L^2$$

We can then use trigonometry to give us the different formulae, dependent on the values that we have been given:

$$\cos \emptyset = \frac{V_R}{V_S} \qquad \sin \emptyset = \frac{V_L}{V_S} \qquad \tan \emptyset = \frac{V_L}{V_R}$$

Example 1

A coil of 0.15H is connected in series with a 50 Ω resistor across a 100 V 50 Hz supply. Calculate the following:

a. The inductive reactance of the coil

b. The impedance of the circuit

c. The circuit current

a. Inductive reactance (X_L)

For inductive reactance, we use the formula

$$X_L = 2 \pi f L \ (\Omega)$$

Inserting the values, this would give us

$$X_L = 2 \times 3.142 \times 50 \times 0.15 \text{ therefore } X_L = 47.13 \ \Omega$$

b. Circuit impedance (Z)

When we have resistance and inductance in series, we calculate the impedance using the following formula:

$$Z^2 = R^2 + X_L^2$$

which becomes

$$Z = \sqrt{R^2 + X_L^2}$$

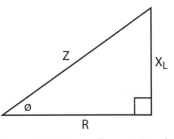

Figure 1.11   Impedance triangle

In the case of the first formula, isn't this the same as Pythagoras' Theorem for a right-angled triangle ($A^2 = B^2 + C^2$)? We therefore sometimes refer to this as the impedance triangle and it can be drawn for this type of circuit, as shown. Here, the angle (Ø) between sides R and Z is the same as the phase angle between sides R and Z, is the same as the phase angle between current and voltage.

If we therefore apply some trigonometry, the following applies:

$$\cos \emptyset = \frac{R}{Z} \qquad \sin \emptyset = \frac{X_L}{Z} \qquad \tan \emptyset = \frac{X_L}{R}$$

However, using our formula

$$Z = \sqrt{R^2 + X_L^2}$$

then $Z = \sqrt{50^2 + 47.1^2}$ therefore $Z = \mathbf{68.69 \ \Omega}$

c. **Circuit current (I)**

As we are referring to the total opposition to current, we use the formula

$$I = \frac{V}{R} = \frac{100}{68.69} = \textbf{1.46 A}$$

Example 2

A coil of 0.159H is connected in series with a 100 Ω resistor across a 230 V 50 Hz supply. Calculate the following:

a. **The inductive reactance of the coil**

b. **The circuit impedance**

c. **The circuit current**

d. **The p.d. across each component**

e. **The circuit phase angle**

a. **Inductive reactance (X_L)**

$$X_L = 2\pi f L \quad \text{therefore} \quad X_L = 2 \times 3.142 \times 50 \times 0.159 = \textbf{50 }\Omega$$

b. **Circuit impedance (Z)**

$$Z = \sqrt{R^2 + X_L^2} \quad \text{therefore} \quad Z = \sqrt{100^2 + 50^2} = \textbf{111.8 }\Omega$$

c. **Circuit current (I)**

$$I = \frac{V}{Z} \quad \text{therefore} \quad I = \frac{230}{111.8} = \textbf{2.06 A}$$

d. **The p.d. across each component (V)**

$$V_R = I \times R \quad \text{therefore} \quad V = 2.06 \times 100 = \textbf{206 V}$$

$$V_L = I \times X_L \quad \text{therefore} \quad V = 2.06 \times 50 = \textbf{103 V}$$

e. **Circuit phase angle (Ø)**

Using our right-angled triangle:

$$\tan Ø = \frac{V_L}{V_R} \quad \text{therefore} \quad \tan Ø = \frac{103}{206} = \textbf{0.5}$$

If you then enter 0.5 into your calculator and press the INV key, then press TAN, you should get the number 26.6.

Therefore, the current is lagging voltage by **26.6°**

Resistance and capacitance in series (RC)

Consider the following diagram.

Figure 1.12   A resistor connected in series with a capacitor, fed from an a.c. supply

Here a resistor is connected in series with a capacitor and fed from an a.c. supply. Once again, in a series circuit the current (I) will be common to both the resistor and the capacitor, causing voltage to drop (p.d.) V_R across the resistor and V_C across the capacitor.

As with the Resistance/Inductance (RL) circuit previously, we can take current as the reference phasor. Similarly, the voltage across the resistor will be in phase with that current. Also, if you remember, in a capacitive circuit the current leads the voltage by 90°. Therefore we can say that the voltage across the capacitor will be lagging the current. As before, we can now calculate the supply voltage (V_S) by completion of the parallelograms as follows:

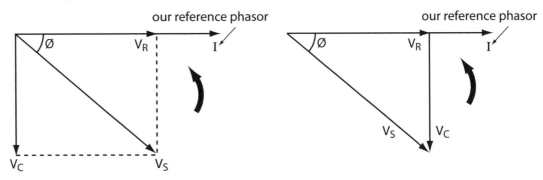

Figure 1.13   Phasor diagrams

As with the inductor, we can apply Pythagoras' Theorem and trigonometry to give us the following formulae:

$$V_S^2 = V_R^2 + V_C^2$$

$$\cos \varnothing = \frac{V_R}{V_S} \qquad \sin \varnothing = \frac{V_C}{V_S} \qquad \tan \varnothing = \frac{V_C}{V_R}$$

Here's an example:

Example

A capacitor of 15.9 μF and a 100 Ω resistor are connected in series across a 230 V 50 Hz supply. Calculate the following:

a. **The circuit impedance**

b. **The circuit current**

c. **The p.d. across each component**

d. **The circuit phase angle**

a. **Circuit impedance (Z)**

To be able to find the impedance we must first find the capacitive reactance.

$$X_C = \frac{1}{2\pi fC}$$

however, as the capacitor value is given in μF, we use $X_C = \frac{10^6}{2\pi fC}$

This gives us : $X_C = \dfrac{10^6}{2 \times 3.142 \times 50 \times 15.9} = \dfrac{10^6}{4995.78} = \mathbf{200\ \Omega}$

When we have resistance and capacitance in series, we use the following formula:

$$Z^2 = R^2 + X_C{}^2 \text{ which becomes } Z = \sqrt{R^2 + X_C{}^2}$$

therefore $\quad Z = \sqrt{100^2 + 200^2} = \sqrt{50\,000} = \mathbf{224\ \Omega}$

b. **Circuit current (I)**

$$I = \frac{V}{Z} \qquad \text{therefore} \quad I = \frac{230}{224} = \mathbf{1.03\ A}$$

c. **The p.d. across each component (V)**

$$V_R = I \times R \qquad \text{therefore} \quad V_R = 1.03 \times 100 = \mathbf{103\ V}$$

$$V_C = I \times X_L \qquad \text{therefore} \quad V_C = 1.03 \times 200 = \mathbf{206\ V}$$

d. **Circuit phase angle (Ø)**

Using our right-angled triangle:

$$\tan \text{Ø} = \frac{V_C}{V_R} \qquad \text{therefore} \quad \tan \text{Ø} = \frac{206}{103} = \mathbf{2}$$

If you then enter 2 into your calculator and press the INV key, then press TAN, you should get the number 63.4.

Therefore, the current is leading voltage by **63.4°**

Resistance, inductance and capacitance in series (RLC)

Consider the following diagram:

Figure 1.14   Resistor connected in series with an inductor and capacitor, fed from an a.c. supply.

Here we have a resistor connected in series with an inductor and a capacitor then fed from an a.c. supply. This is often referred to as an RLC circuit or a general series circuit. Again, as we have a series circuit, the current (I) will be common to all three components, causing a voltage drop (p.d.) V_R across the resistor, V_L across the inductor and V_C across the capacitor.

Here V_R will be in phase with the current, V_L will lead the current by 90° (because the current lags the voltage) and V_C will lag the current by 90° (because current leads the voltage in a capacitive circuit). Because V_L and V_C are in opposition to each other (one leads and one lags), the actual effect will be the difference between their values, subtracting the smaller from the larger. We can once again calculate V_S by completing a parallelogram as follows.

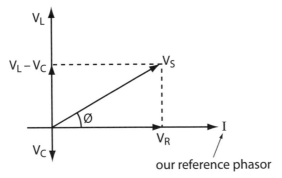

Figure 1.15   Phasor parallelogram

As before, we can now apply Pythagoras' Theorem and trigonometry to give us the following formulae depending on whether V_L or V_C is the larger:

$$V_S^2 = V_R^2 + (V_L - V_C)^2 \quad \text{or} \quad V_S^2 = V_R^2 + (V_C - V_L)^2$$

and because I is the same for each component we get:

$$Z = \sqrt{R^2 + (X_L - X_C)^2} \quad \text{or} \quad Z = \sqrt{R^2 + (X_C - X_L)^2}$$

and finally: $\cos \emptyset = \dfrac{V_R}{V_S} \quad \sin \emptyset = \dfrac{V_L - V_C}{V_S} \quad \tan \emptyset = \dfrac{V_L - V_C}{V_R}$

Example

A resistor of 5 Ω is connected in series with an inductor of 0.02H and a capacitor of 150 μF across a 250 V 50 Hz supply. Calculate the following.

a. The impedance

b. The supply current

c. The power factor

a. Impedance (Z)

In order to find the impedance, we must first find out the relevant values of reactance.

Therefore: $X_L = 2\pi f L = 2 \times 3.142 \times 50 \times 0.02 = \mathbf{6.28\ \Omega}$

$X_C = \dfrac{1}{2\pi f C}$ allowing for microfarads

$= \dfrac{10^6}{2 \times 3.142 \times 50 \times 150} = \mathbf{21.2\ \Omega}$

If you remember, we said that the effect of inductance and capacitance together in series would be the difference between their values. Consequently, this means that the resulting reactance (X) will be found as follows:

$X = X_L - X_C$ or, in this case because X_C is the larger: $X = X_C - X_L$

and therefore: $X = 21.2 - 6.28 = \mathbf{14.92\ \Omega}$

We can now use the impedance formula as follows:

$Z = \sqrt{R^2 + (X_L - X_C)^2}$ or $Z = \sqrt{R^2 + (X_C - X_L)^2}$

which gives us

$Z = \sqrt{5^2 + 14.92^2} = \mathbf{15.74\ \Omega}$

Note: X_C is greater than X_L. Therefore we subtract X_L from X_C. Had X_L been the higher, then the reverse would be true. Also, as capacitive reactance is highest, the circuit current will lead the voltage. Had the inductive reactance been the higher, then the current would lag the voltage.

b. Supply current (I)

$I = \dfrac{V}{Z} = \dfrac{230}{15.74} = 14.6\ A$

c. Power factor (Ø)

$\cos \varnothing = \dfrac{R}{Z} = \dfrac{5}{15.74} = 0.32$

Therefore PF = **0.32 leading**

Resistance, inductance and capacitance in parallel

Consider the following diagram:

Figure 1.16 Resistor, capacitor and inductor in parallel connected to an a.c. supply

There can obviously be any combination of the above components in parallel. However, to demonstrate the principles involved, we will look at all three connected across and a.c. supply.

As we have a parallel circuit, the voltage (V_S) will be common to all branches of the circuit. Consequently, when we draw our parallelogram we will use voltage as the reference phasor.

In this type of circuit, the current through the resistor will be in phase with the voltage, the current through the inductor will lag the voltage by 90° and the current through the capacitor will lead the voltage by 90°.

Normally, when we carry out calculations for parallel circuits, it is easier to treat each branch as being a separate series circuit. We then draw to scale each of the respective currents and their relationships to our reference phasor, which is voltage.

As with voltage V_L and V_S in the RLC series circuit, the current through the inductor (I_L) and the current through the capacitor (I_C) are in complete opposition to each other. Therefore, the actual effect will be the difference between their two values. We calculate this value by the completion of our parallelogram. But, bear in mind that the bigger value (I_C or I_L) will determine whether the current ends up leading or lagging.

If we use the diagram as the basis for our example, calculate the circuit current and its phase angle relative to the voltage.

To calculate the circuit current and its phase angle relative to the voltage, we must first find the current through each branch:

$$I_R = \frac{V}{R} = \frac{230}{50} = \textbf{4.6 A}$$

$$X_L = 2\pi fL = 2 \times 3.142 \times 50 \times 0.4 = \textbf{126 } \Omega$$

Therefore $\quad I_L = \dfrac{V}{X_L} = \dfrac{230}{126} = \textbf{1.8 A}$

$$X_C = \frac{10^6}{2\pi fC} = \frac{10^6}{2 \times 3.142 \times 50 \times 60} = \textbf{53 } \Omega$$

Therefore $I_C = \dfrac{V}{X_C} = \dfrac{230}{53} = \textbf{4.3 A}$

The actual effect will be $I_C - I_L$ which gives $4.3 - 1.8 = \textbf{2.5 A}$

Now add this to I_R by completing the scale drawing at Figure 1.17. This gives a current of **5.2 A** that is leading voltage by an angle of **28°**.

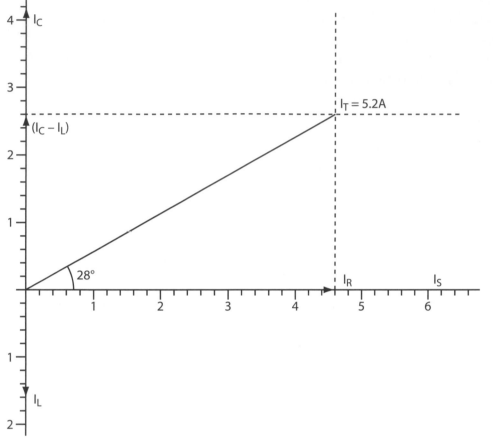

Figure 1.17   Scale drawing

Power in an a.c. circuit

If we were to try to push something against a resistance, we would get hot and bothered as we use up energy in completing the task. When current flows through a resistor, a similar thing happens in that power is the rate of using up energy – in other words, the amount of energy that was used in a certain time.

If a resistor (R) has a current (I) flowing through it for a certain time (t), then the power (energy being used per second) given in watts, can be calculated by the formula:

$P = I^2 \times R$

This power will be dissipated in the resistor as heat and reflects the average power in terms of the r.m.s. values of voltage and current.

What we are effectively saying her is that the average power in a resistive circuit (one which is non-reactive, i.e. doesn't possess inductance or capacitance) can be found by the product of the readings of an ammeter and a voltmeter.

In other words, in the resistive circuit, the power (energy used per second) is associated with that energy being transferred from the medium of electricity into another medium, such as light (filament lamp) or heat (electric fire/kettle). We call this type of power the active power. When we look at the capacitive circuit, we find that current flows to the capacitor, but we have no power.

Look at this wave diagram for voltage (Figure 1.18).

During the first period of the cycle, the voltage is increasing and this provides the energy to charge the capacitor. However, in the second period of the cycle, the voltage is decreasing and therefore the capacitor discharges, returning its energy back to the circuit as it does so. The same is also true of the third and fourth periods.

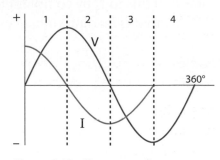

Figure 1.18　Sine wave diagram

This exchange of energy means that we have voltage and current, but no average power and therefore no heating effect. This means that our previous formula ($P = I^2 \times R$) is no longer useful.

We therefore say that the result of voltage and current in this type of circuit is called reactive power and we express this in reactive voltamperes (VAr). Equally, we say that the current in a capacitive circuit, where there is no resistance and no dissipation of energy, is called **reactive current**.

When we come to the inductive circuit we have a similar position. This time, as voltage increases during the first period of the cycle, the energy is stored as a magnetic field in the inductor. This energy will then be fed back into the circuit during the second period of the cycle as the voltage decreases and the magnetic field collapses. In other words, once again the exchange of energy produces no average power (energy used per second).

Circuits are likely to comprise combinations of resistance, inductance and capacitance. In a circuit, which has resistance and reactance, there will be a phase angle between the voltage and current. This relationship has relevance, as power will only be expended in the resistive part of the circuit.

Let's now look at this relationship via a phasor diagram, for a circuit containing resistance and capacitance, remembering that for this type of circuit the current will lead the voltage.

This circuit has two components (resistance and capacitance), so we've drawn two current phasors. In reality, as we already know, these are not currents that will actually flow, but their phasor sum will be the actual current in the circuit I_p (actual).

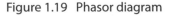

Figure 1.19　Phasor diagram

We also said that in a resistive circuit the voltage and current are in phase, and therefore this section of the current has been represented by the phasor I_p(active). This part of the current is in the active section and we therefore refer to this as the active current.

We then show that part of the current in the reactive section (capacitor) as leading the voltage by 90° and this has been represented by the phasor I_Q(reactive). As stated previously, we refer to this as the reactive current.

We also know that for a resistive circuit, we calculate the power by multiplying together the r.m.s. values of voltage and current ($V \times I$).

Logically, as we know that no power is consumed in the reactive section of the circuit, we can therefore calculate the power in the circuit by multiplying together the r.m.s. value of voltage and the value of current, which is in phase with it (I_p(active)). This would give us the formula:

$$P = V \times I_p\text{(active)}$$

But as the actual current will be affected by the reactive current and therefore the phase angle (cos Ø), our formula becomes:

$$P = V \times I_p\text{(active)} \times \cos Ø$$

If Ip(active) is zero then Ø = 90° and cos Ø = 0. Therefore P = 0.

We have now established that it is possible in an a.c. circuit for current to flow, but no power to exist.

We also say that the product of voltage and current is power given in watts. However, it would be fair to say that this is not the actual power of the circuit.

The actual (true) power of the circuit has to take on board the effect of the phase angle (cos Ø), the ratio of these two statements being the power factor. In other words:

$$\text{Power Factor (cos Ø)} = \frac{\text{True power (P)}}{\text{Apparent power (S)}}$$

$$= \frac{V \times I_p\text{(actual)} \times \cos Ø}{V \times I_p\text{(actual)}} = \frac{\text{watts}}{\text{volt-amperes}}$$

To summarise, what we perceive to be the power of a circuit (the apparent power) can also be the true power, as long as we have a unity power factor (1.0). However, as long as we have a phase angle, then we have a difference between apparent power and reality (true power). This difference is the power factor (a value less than unity).

In reality we will use a wattmeter to measure the true power and a voltmeter and ammeter to measure the apparent power.

The power triangle

Figure 1.20

We can use Pythagoras' Theorem to help us calculate the different power components within a circuit.

We do so by using Pythagoras' formula as follows:

$$(VA)^2 = (W)^2 + (VAr)^2$$

This is then applied to Figure 1.20, where we have shown both inductive and capacitive conditions.

Example

A resistor of 15 Ω has been connected in series with a capacitor of reactance 30 Ω. If they are connected across a 230 V supply, establish both by calculation and by drawing a scaled power triangle, the following:

(a)   the apparent power   (b)   the true power

(c)   the reactive power   (d)   the power factor.

By calculation

In order to establish the elements of power, we must first find the current. To do this we need to find the impedance of the circuit. Therefore:

$$Z = \sqrt{R^2 + X_C^2} = \sqrt{15^2 + 30^2} = \sqrt{225 + 900} = \sqrt{1125}$$

Therefore $Z = 33.5\,\Omega$

$$I = \frac{V}{Z} = \frac{230}{35.5} = 6.9\,A$$

$$Pf(\cos)\,\varnothing = \frac{R}{Z} = \frac{15}{33.5} = 0.45$$

True power $= I \times V \times Pf$

$\qquad\qquad = 6.9 \times 230 \times 0.45$

$\qquad\qquad = 714.15\,W\ \ or\ \ 0.714\,kW$

Apparent power $= I \times V$

$\qquad\qquad\quad = 6.9 \times 230$

$\qquad\qquad\quad = 1587\,VA\ \ or\ \ 1.587\,kVA$

Reactive power $= \sin\varnothing \times$ apparent power

(graph labels: 1.59 kVA, 1.41 kVAr, 63.2°, 714 kW, 1)

However, we do not know the value of sin Ø, we must therefore convert cos Ø to an angle and then find the sine of that angle.

cos Ø = 0.45   INV cos of 0.45 = 63.2

sine of 63.2° = 0.89

Therefore Reactive power = sin Ø × apparent power
$$= 0.89 \times 1587$$
$$= 1412.43 \, \text{VAr} \quad \text{or} \quad 1.412 \, \text{kVAr}$$

Three-phase power

As we have seen in the previous section, we can find the power in a single-phase a.c. circuit by using the following formula:

Power = $V \times I \times \cos Ø$

Logically, you might assume that for three phase we could multiply this formula by three. Although this is not far from the truth we must remember that this could only apply where we have a balanced three-phase load. We can therefore say, that for any three-phase balanced load, the formula to establish power is:

Power = $\sqrt{3} \times (V_L \times I_L \times \cos Ø)$

However, in the case of an unbalanced load, we need to calculate the power for each separate section and then add them together to get total power.

This was all covered in Electrical Installations Book 1, Chapter 3 and you may like to refresh your memory before working through the three examples that follow.

Example 1

A balanced load of 10 Ω per phase is star connected and supplied with 400 V 50 Hz at unity power factor. Calculate the following:

a. **Phase voltage**

b. **Line current**

c. **Total power consumed**

a. **Phase voltage**

$$V_L = \sqrt{3} \times V_P$$

Therefore by transposition

$$V_P = \frac{V_L}{\sqrt{3}} = \frac{400}{1.732} = \textbf{231 V}$$

b. **Line current**

$$I_L = I_P$$

and therefore

$$I_P = \frac{V_P}{R_P} = \frac{231}{10} = \textbf{23.1 A}$$

c. **Total power**

In a balanced system

$$\text{Power} = \sqrt{3} \times V_L \times I_L \times \cos Ø$$
$$\text{Power} = 1.732 \times 400 \times 23.1 \times 1 = \textbf{16 kW}$$

Example 2

Three coils of resistance 40 Ω and inductive reactance 30 Ω are connected in delta to a 400 V 50 Hz three-phase supply. Calculate the following:

a. The current in each coil

b. Line current

c. Total power

a. Current in each coil

We must first find the impedance of each coil:

$$Z = \sqrt{R^2 + X_L{}^2} = \sqrt{40^2 + 30^2} = \sqrt{2500} = 50\ \Omega$$

The current in each coil (I_p) can then be found by applying Ohm's Law:

This gives $I_P = \dfrac{V}{Z} = \dfrac{400}{50} = \textbf{8 A}$

b. Line current

For a delta connected system

$$I_L = \sqrt{3} \times I_P$$

Therefore $I_L = 1.732 \times 8 = \textbf{13.86 A}$

c. Total power

We must first find the power factor using the formula:

$$\cos \emptyset = \frac{R}{Z}$$

This gives us $\cos \emptyset = \dfrac{40}{50} = 0.8$

And for a delta system $V_L = V_P$

Therefore we can now use the power formula of :

$$P = \sqrt{3} \times V_L \times I_L \times \cos \emptyset$$

$$P = 1.732 \times 400 \times 13.86 \times 0.8$$

$$P = \textbf{7682 W} \text{ or } \textbf{7.682 kW}$$

Example 3

A small industrial estate is fed by a 400 V, three-phase, 4-wire TN-S system. On the estate there are three factories connected to the system as follows:

Factory A taking 50 kW at unity power factor

Factory B taking 80 kVA at 0.6 lagging power factor

Factory C taking 40 kVA at 0.7 leading power factor

Calculate the overall kW, kVA, kVar and power factor for the system.

To clarify, we are trying to find values of P (true power), S (apparent Power) and Q (reactive power). First, we need to work out the situations for each factory.

Factory A

We know that power factor

$$\cos \varnothing = \frac{\text{True power (P)}}{\text{Apparent power (S)}}$$

We also know that the power factor is 1.0 and that P = 50 kW

Therefore, by transposition:

$$S = \frac{P}{\cos \varnothing} = \frac{50}{1} = 50 \text{ kVA}$$

And with unity power factor for Factory A, Q = 0

Factory B

Using the same logic, we need to find true power and reactive power.

Therefore P = cos ∅ x S = 0.6 × 80 kW = 48 kW

Reactive component (Q) = S × sin ∅ = 80 × 0.8 = 64 kVAr

Factory C

$$P = S \times \cos \varnothing = 40 \text{ kW} \times 0.7 = 28 \text{ kW}$$

$$Q = S \times \sin \varnothing = 40 \times 0.714 = 28.6 \text{ kVAr}$$

We can now find the **total kW** by addition: 50 + 48 + 28 = **126 kW**

We can find the **total kVAr** as the difference between the reactive power components, the larger one, Factory B, lagging and the smaller one, Factory C, leading:

64 kVAr – 28.6 kVAr = **35.4 kVAr**

We can use Pythagoras' Theorem to find the **total kVA**

$$S = \sqrt{P^2 + Q^2} = \sqrt{126^2 + 35.4^2} = \textbf{131 kVA}$$

Consequently, the overall power factor will be:

$$\cos \varnothing = \frac{P}{S} = \frac{126}{131} = \textbf{0.96 lagging}$$

Transformers

The **transformer** is one of the most widely used pieces of electrical equipment and can be found in situations such as electricity distribution, construction work and electronic equipment. Its purpose, as the name implies, is to transform something – the something in this case being the voltage, which can enter the transformer at one level (input) and leave at another (output).

When the output voltage is higher than the input voltage we say that we have a step-up transformer and when the output voltage is lower than the input, we say that we have a step-down transformer.

In this section we will be looking at the following areas:

● mutual inductance

● transformer types

● step-up and step-down transformers.

Mutual inductance

In their operation, transformers make use of an action known as **mutual inductance**. We have looked at inductance in another section. We will revisit the subject here to help you try to understand the concept clearly. Let us look at the following situation. Two coils, primary and secondary, are placed side by side,

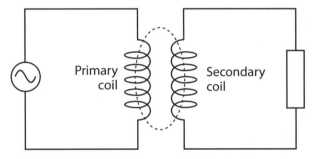

Figure 1.21   Mutual inductance

but not touching each other. The primary coil has been connected to an a.c. supply and the secondary coil is connected to a load, such as a resistor.

If we now allow current to flow in the primary coil, we know that current flow will create a magnetic field. Therefore, as the current in the primary coil increases up to its maximum value, it creates a changing magnetic flux and, as long as there is a changing magnetic flux, there will be an e.m.f. 'induced' into the secondary coil, which would then start flowing through the load.

This effect, where an alternating e.m.f. in one coil causes an alternating e.m.f. in another coil, is known as mutual inductance.

In transformers we are only really interested in mutual inductance, and it is the rising and falling a.c. current that causes the change of magnetic flux. In other words, **we really need an a.c. supply to allow transformers to operate correctly**.

If the two coils were now wound on an iron core, we would find that the level of magnetic flux is increased and consequently the level of mutual inductance is also increased.

Did you know?

There is something called self-inductance. This is when a conductor that is formed into a coil, carries an alternating current, the magnetic field created around the coil grows and collapses with the changing current, this will induce into the coils an e.m.f. voltage that is always in opposition to the voltage that creates it

Transformer types

Core-type transformers

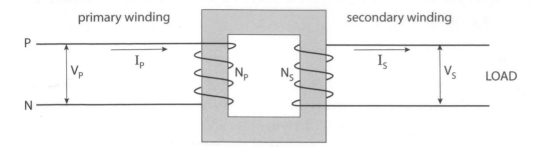

Figure 1.22   Double wound, core type transformer

In Figure 1.22, the supply is wound on one side of the iron core (primary winding) and the output is wound on the other (secondary winding). In other words, double wound means that there is more than one winding.

The number of turns in each winding will affect the induced e.m.f., with the number of turns in the primary being referred to as (Np), and those in the secondary referred to as (Ns). We call this the turns ratio. When voltage (Vp) is applied to the primary winding, it will cause a changing magnetic flux to circulate in the core. This changing flux will cause an e.m.f. (Vs) to be induced in the secondary winding.

Assuming that we have no losses or leakage (i.e. 100 per cent efficient), then power input will equal power output and the ratio between the primary and secondary sides of the transformer can be expressed as follows:

$$\frac{V_P}{V_S} = \frac{N_P}{N_S} = \frac{I_S}{I_P}$$

(where I_P represents the current in the primary winding and I_S the current in the secondary winding).

As we can see from Figure 1.22, a transformer has no moving parts. Consequently, provided that the following general statements apply, it ends up being a very efficient piece of equipment:

- Transformers use laminated (layered) steel cores, not solid metal. In a solid metal core 'eddy currents' are induced which cause heating and power losses.
- Laminated cores, where each lamination is insulated, help to reduce this effect.
- Soft iron with high magnetic properties is used for the core.
- Windings are made from insulated, low resistance conductors. This prevents short circuits occurring either within the windings, or to the core.

The losses that occur in transformers can normally be classed under the following categories.

Remember

No transformer can be 100 per cent efficient. There will always be power losses

Copper losses

Although windings should be made from low resistance conductors, the resistance of the windings will cause the currents passing through them to create a heating effect and subsequent power loss. This power loss can be calculated using the formula:

$$P_c = I^2 \times R \text{ watts}$$

Iron losses

> **Definition**
>
> **Hysteresis** – a generic term meaning a lag in the effect of a change of force

These losses take place in the magnetic core of the transformer. They are normally caused by eddy currents (small currents which circulate inside the laminated core of the transformer) and **hysteresis**. To demonstrate let's say that you push on some material and it bends. When you stop pushing, does it return to its original shape? If it doesn't the material is demonstrating hysteresis. Let's look at this in context.

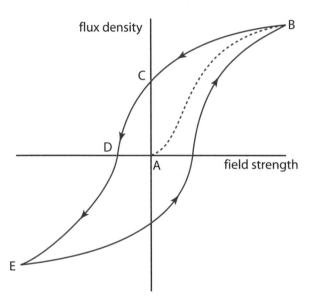

Figure 1.23   Hysteresis loop diagram

Figure 1.23 shows the effect within ferromagnetic materials of hysteresis. Starting with an unmagnetised material at point A; and here both field strength and flux density are zero. The field strength increases in the positive direction and the flux begins to grow along the dotted path until we reach saturation at point B. This is called the initial magnetisation curve.

If the field strength is now relaxed, instead of retracing the initial magnetisation curve, the flux falls more slowly. In fact, even when the applied field has returned to zero, there will still be a degree of flux density (known as the **remanence**) at point C. To force the flux to go back to zero (point D), we have to reverse the applied field. The field strength here that is necessary to drive the field back to zero is known as the **coercivity**. We can then continue reversing the field to get to point E, and so on. This is known as the hysteresis loop.

As we have already said, we can help reduce eddy currents by using a laminated core construction. We can also help to reduce hysteresis by adding silicon to the iron from which the transformer core is made.

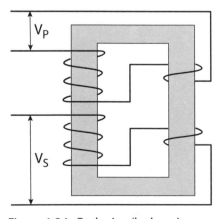

Figure 1.24   Reducing 'leakage'

The version of the double wound transformer that we have looked at so far makes the principle of operation easier to understand.

However, this arrangement is not very efficient, as some of the magnetic flux being produced by the primary winding will not react with the secondary winding and is often referred to as 'leakage'. We can help to reduce this leakage by splitting each winding across the sides of the core (see Figure 1.24).

Shell-type transformers

We can reduce the magnetic flux leakage a bit more, by using a shell-type transformer.

In the shell-type transformer, both windings are wound onto the central leg of the transformer and the two outer legs are then used to provide parallel paths for the magnetic flux.

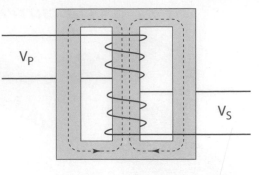

Figure 1.25   Shell-type transformer

The autotransformer

The autotransformer uses the principle of 'tapped' windings in its operation. Remember that the ratio of input voltage against output voltage will depend upon the number of primary winding turns and secondary winding turns (the turns ratio). But what if we want more than one output voltage?

Some devices are supplied with the capability of providing this, such as small transformers for calculators, musical instruments or doorbell systems. Tapped connections are the normal means by which this is achieved. A tapped winding means that we have made a connection to the winding and then brought this connection out to a terminal. Now, by connecting between the different terminals, we can control the number of turns that will appear in that winding and we can therefore provide a range of output voltages.

An autotransformer has only one tapped winding and the position of the tapping on that winding will dictate the output voltage.

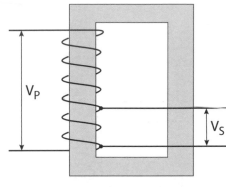

Figure 1.26   Autotransformer

One of the advantages of the autotransformer is that, because it only has one winding, it is more economical to manufacture. However, on the down side, we have made a physical connection to the winding. Therefore, if the winding ever became broken between the two tapping points, then the transformer would not work and the input voltage would appear on the output terminals. This would then present a real hazard.

Instrument transformers

Instrument transformers are used in conjunction with measuring instruments because it would be very difficult and expensive to design normal instruments to measure the high current and voltage that we find in certain power systems. We therefore have two types of instrument transformer, both being double wound.

The current transformer

Current transformer

The current transformer (c.t.) normally has very few turns on its primary winding so that it does not affect the circuit to be measured, with the actual meter connected across the secondary winding.

Care must be taken when using a c.t. Never open the secondary winding while the primary is 'carrying' the main current. If this happened, a high voltage would be induced into the secondary winding. Apart from the obvious danger, this heat build up could cause the insulation on the c.t. to break down.

The voltage transformer

This is very similar to our standard power transformer, in that it is used to reduce the system voltage. The primary winding is connected across the voltage that we want to measure and the meter is connected across the secondary winding.

Step-up and step-down transformers

Step-up transformers

120 V a.c. Supply
10 amps

240 V a.c. Output
5 amps

FIgure 1.27   Step-up transformer

A step-up transformer is used when it is desirable to step voltage up in value.

The primary coil has fewer turns than the secondary coil. We already know that the number of turns in a transformer is given as a ratio. When the primary has fewer turns than the secondary, voltage and impedance are stepped up. In the circuit shown, voltage is stepped up from 120 V a.c. to 240 V a.c. Since impedance is also stepped up, current is stepped down from 10 amps to 5 amps.

Step-down transformers

240 V a.c. Supply
5 amps

120 V a.c. Output
10 amps

FIgure 1.28   Step-down transformer

A step-down transformer is used when it is desirable to step voltage down in value.

The primary coil has more turns than the secondary coil. The step-down ratio is 2:1. Since the voltage and impedance are stepped down, the current is stepped up in this case to 10 A.

Safety isolating transformer

Another use of a transformer is to isolate the secondary output from the supply. In a bathroom, the shower socket should be supplied from a 1:1 safety isolating transformer. The output of the transformer has no connection to earth thereby ensuring that output from the transformer is totally isolated from the supply.

In areas of installations where there is an increased risk of electric shock, the voltage is reduced to less than 50 V and is supplied from a safety isolating transformer; we know this as **Separate Extra Low Voltage** (**SELV**) supply.

Transformer ratings

Transformers are rated in kVA (kilovolt-amps). This rating is used rather than watts because loads are not purely resistive. Only resistive loads are measured in watts.

The kVA rating determines the current that a transformer can deliver to its load without overheating. Given volts and amps, kVA can be calculated. Given kVA and volts, amps can be calculated.

The kVA rating of a transformer is the same for both the primary and the secondary. At this point let us try some examples.

Example 1

A transformer having a turns ratio of 2:7 is connected to a 230 V supply. Calculate the output voltage.

When we give transformer ratios, we give them in the order primary then secondary. Therefore in this example we are saying that for every two windings on the primary winding, there are seven on the secondary. If we therefore use our formula:

$$\frac{V_P}{V_S} = \frac{N_P}{N_S}$$

We should now transpose this to get:

$$V_S = \frac{V_P \times N_S}{N_P}$$

However, we do not know the exact number of turns involved. But do we need to, if we know the ratio? Let us find out.

The ratio is 2:7, meaning for every 2 turns on the primary, there will be 7 turns on the secondary. Therefore, if we had 6 turns on the primary, this would give us 21 turns on the secondary, but the ratio of the two has not changed. It is still for every 2 on the primary we are getting 7 on the secondary.

This means we can just insert the ratio rather than the individual number of turns into our formula:

$$V_s = \frac{V_P \times N_S}{N_P} = \frac{230 \times 7}{2} = \textbf{805V}$$

Now, to prove our point about ratios, let us say that we know the number of turns in the windings to be 6 in the primary and 21 in the secondary (which is still giving us a 2:7 ratio). If we now apply this to our formula we get:

$$V_s = \frac{V_P \times N_S}{N_P} = \frac{230 \times 21}{6} = \textbf{805V}$$

The same answer.

Example 2

A single phase transformer, with 2000 primary turns and 500 secondary turns, is fed from a 230V a.c. supply. Find:

(a) the secondary voltage

(b) the volts per turn.

Secondary voltage

$$\frac{V_P}{V_S} = \frac{N_P}{N_S}$$

Using transposition, re-arrange the formula to give:

$$V_s = \frac{V_P \times N_S}{N_P}$$

$$Vs = \frac{230 \times 500}{2000} = \frac{115,000}{2000} = \textbf{57.5V}$$

Volts per turn

This is the relationship between the volts in a winding and the number of turns in that winding. To find volts per turn, we simply divide the voltage by the number of turns.

Therefore, in the primary:

$$\frac{V_P}{N_P} = \frac{230}{2000} = \textbf{0.115 volts per turn}$$

In the secondary

$$\frac{V_S}{N_S} = \frac{57.5}{500} = \textbf{0.115 volts per turn}$$

Example 3

A single phase transformer is being used to supply a trace heating system. The transformer is fed from a 230V 50 Hz a.c. supply and needs to provide an output voltage of 25 V. If the secondary current is 150 A and the secondary winding has 50 turns, find:

(a)   the output kVA of the transformer

(b)   the number of primary turns

(c)   the primary current

(d)   the volts per turn.

The output kVA

$$\text{kVA} = \frac{\text{volts} \times \text{amperes}}{1000} = \frac{V_S \times I_S}{1000} = \frac{25 \times 150}{1000} = \textbf{3.75 kVA}$$

The number of primary turns

If: $\dfrac{V_P}{V_S} = \dfrac{N_P}{N_S}$ then by transposition

$$N_P = \frac{V_P \times N_S}{V_S} = \frac{250 \times 50}{25} = \textbf{460 turns}$$

The primary current

If: $\dfrac{V_P}{V_S} = \dfrac{I_S}{I_P}$ then by transposition

$$I_P = \frac{V_S \times I_S}{V_P} = \frac{25 \times 150}{230} = \textbf{16 A}$$

The volts per turn

In the primary: $\dfrac{V_P}{N_P} = \dfrac{230}{460} = \textbf{0.5 volts per turn}$

In the secondary: $\dfrac{V_S}{N_S} = \dfrac{25}{50} = \textbf{0.5 volts per turn}$

Example 4

A step-down transformer, having a ratio of 2:1, has an 800 turn primary winding and is fed from a 400 V a.c. supply. The output from the secondary is 200 V and this feeds a load of 20 Ω resistance. Calculate:

(a) the power in the primary winding

(b) the power in the secondary winding.

Well, we know that the formula for power is: $P = V \times I$

And we also know that we can use Ohm's Law to find current: $I = \dfrac{V}{R}$

Therefore, if we insert the values that we have, we can establish the current in the secondary winding:

$$I_s = \frac{V_s}{R_s} = \frac{200}{20} = \textbf{10 A}$$

Now that we know the current in the secondary winding, we can use the power formula to find the power generated in the secondary winding:

$$P = V \times I = 200 \times 10 = 2000 \text{ watts} = \textbf{2 kW}$$

We now need to find the current in the primary winding. To do this we can use the formula:

$$\frac{V_P}{V_S} = \frac{I_S}{I_P}$$

However, we need to transpose the formula to find I_p. This would give us:

$$I_p = \frac{I_S \times V_S}{V_P}$$

Which, if we now insert the known values to, gives us:

$$I_p = \frac{I_S \times V_S}{V_P} = \frac{10 \times 200}{400} = \frac{2000}{400} = \textbf{5 A}$$

Now that we know the current in the primary winding, we can again use the power formula to find the power generated in the secondary winding:

$$P = V \times I = 400 \times 5 = 2000 \text{ watts} = \textbf{2 kW}$$

Instruments and measurement

As electricians we are often responsible for the measurement of different electrical quantities. Some must be measured as part of the inspection and testing of an installation (e.g. Insulation Resistance). However, of the remainder, the most common quantities that we are likely to come across are shown in Table 1.1.

Property	Instrument
Current	Ammeter
Voltage	Voltmeter
Resistance	Ohmmeter
Power	Wattmeter

Table 1.1   Electrical quantities

In this section we will be looking at the following areas:

- measuring current (ammeter)
- measuring voltage (voltmeter)
- measuring resistance (ohmmeter)
- using an ammeter and voltmeter
- loading errors
- meter displays (digital and analogue).

We will look at the different instruments that we use to measure these quantities in the following section.

But before we ever use a meter, we must always ask ourselves:

- Have we chosen the correct instrument?
- Is it working correctly?
- Has it been set to the correct scale?
- How should it be connected?

Measuring current (ammeter)

An ammeter measures current by series connections. Although we know that we can use a multi-meter on site, the actual name for an instrument that measures current is an ammeter. Ammeters are connected in series so that the current to be measured passes through them. The circuit diagram in Figure 1.29 illustrates this. Consequently, they need to have a very low resistance or they would give a false reading.

If we now look at the same circuit, but use a correctly set multi-meter, the connections would look as in Figure 1.30.

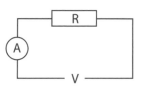

Figure 1.29   Ammeter in circuit

Figure 1.30   Multi-meter in circuit

Measuring voltage (voltmeter)

Figure 1.31 Voltmeter in circuit

On site or for general purposes, we can use a multi-meter to measure voltage, but the actual device used for measuring voltage is called a voltmeter. It measures the potential difference between two points (for instance, across the two connections of a resistor). The voltmeter must be connected in parallel across the load or circuit to be measured as shown in Figure 1.31.

If we now look at the same circuit, but use our correctly set multi-meter, the connections would look as in Figure 1.32.

The internal resistance of a voltmeter must be very high if we wish to get accurate readings.

Figure 1.32 Multi-meter measuring voltage

Measuring resistance (ohmmeter)

Instrument

Resistance under test

Figure 1.33 Ohmmeter

There are many ways in which we can measure resistance. However, in the majority of cases we do so by executing a calculation based upon instrument readings (ammeter/voltmeter) or from other known resistance values. Once again we can use our multi-meter to perform the task, but we refer to the actual meter as being an ohmmeter.

The principle of operation involves the meter having its own internal supply (battery). The current, which then flows through the meter, must be dependent upon the value of the resistance under scrutiny.

However, before we start our measurement, we must ensure that the supply is off and then connect both leads of the meter together and adjust the meter's variable resistor until full-scale deflection (zero) is reached.

The dotted line in Figure 1.33 indicates the circuit under test. All other components are internal to the meter.

Using an ammeter and voltmeter

If we consider the application of Ohm's Law, we know that we can establish resistance by the following formula:

$$R = \frac{V \text{ (voltmeter)}}{I \text{ (ammeter)}}$$

If we therefore connect an ammeter and voltmeter in the circuit, then by using a known supply (battery) we can apply the above formula to the readings that we get.

In practice, we invariably find that we need a long 'wandering' test lead to perform such a test. Where this is the case, we must remember to deduct the value of the test lead from our circuit resistance.

Loading errors

Connecting a measuring instrument to a circuit will invariably have an effect on that circuit. In the case of the average electrical installation, this effect can largely be ignored. But if we were to measure electronic circuits, the effect may be drastic enough to actually destroy the circuit.

We normally refer to these errors introduced by the measuring instrument as loading errors, and below we will look at an example of this.

Let's assume in (a) that we have been asked to measure the voltage across resistor B in the circuit, where both items have a resistance of 100 kΩ. Note that in this arrangement, the potential difference (p.d.) across each resistor will be 115 V.

Our instrument, a voltmeter, has an internal resistance of 100 kΩ and would therefore be connected as in the diagram.

Figure 1.34   Loading errors

In effect, by connecting our meter in this way, we are introducing an extra resistance in parallel into the circuit. We could, therefore, now draw our circuit as in Figure 1.34(b).

Now, if we find the equivalent resistance of the parallel branches:

$$\frac{1}{R_t} = \frac{1}{R_1} + \frac{1}{R_2} = \frac{1+1}{100} = \frac{2}{100}\ k\Omega$$

Therefore:

$$R_t = \frac{100}{2} = 50\ k\Omega$$

And then if we treat the parallel combination as being in series with 100 kΩ resistor A then total resistance:

$$R = R_A + R_t = 100\ k\Omega + 50\ k\Omega = 150\ k\Omega$$

If you applied Ohm's Law again, we would find we now have a current of 0.00153 A (1.53 mA) flowing in the circuit. This in turn would give a p.d. across each resistor of 76 V and not the 115 V we would expect to see.

The problem is caused by the amount of current that is flowing through the meter, and this is normally solved by using meters with a very high resistance (as resistance restricts current). So you now see why we said earlier that voltmeters must have a high internal resistance in order to be accurate.

Analogue meter   Digital meter

Meter displays

There are two types of display, **analogue** and **digital**. Analogue meters have a needle moving around a calibrated scale, whereas digital tend to show results, as numeric values, via a liquid crystal or LED display.

For many years electricians used the analogue meter and indeed it is still common for continuity/insulation resistance testers to be of the analogue style. However, their use is slowly being replaced by the digital meter. Most modern digital meters are based on semiconductors and consequently have very high impedance, making them ideal for accurate readings and good for use with electronic circuits.

Generally speaking, as they have no moving parts they are also more suited to rugged site conditions than the analogue style.

What sort of meter will I need?

Although instruments exist to measure individual electrical quantities, most electricians will find that a cost effective solution is to use a digital multi-meter similar to the one pictured. Meters like this are generally capable of measuring current, resistance and voltage, both in a.c. and d.c. circuits and across a wide range of values.

Is the meter working correctly?

It is important to check that your meter is functioning correctly and physically fit for the purpose. There are commercially available proving units that will help. However, on site it is more normal to use a known supply as the means of proving voltage measurement and the shorting/separating of leads to prove operation.

Measuring power

In a d.c. circuit it is possible to measure the power supplied by using a voltmeter and an ammeter. Then by using the formula **P = V × I** we can arrive at the power in Watts.

However, in an a.c. circuit, this method only produces the apparent power, a figure that will not be accurate unless we have unity power factor. This is because components such as capacitors and inductors will cause the current to lead or lag the voltage.

For a single-phase a.c. circuit we therefore use the formula:

P = V × I × cos Ø

This indicates how we would connect measuring instruments to establish power. For a single-phase resistive load, we can use the ammeter and voltmeter, but for a circuit containing capacitance or inductance we must use a wattmeter.

Figure 1.35 indicates how a wattmeter is connected into a single-phase circuit. Just like Ohm's Law, where to calculate power we need to know the values of voltage and current, a wattmeter also requires values of voltage and current.

The instrument shown in Figure 1.35 has a current coil, indicated between W_1 and W_2, that is wired in series with the load, and a voltage coil, shown between P_1 and P_2, wired in parallel across the supply.

This can be used to measure the power in a single-phase circuit or in a balanced three-phase load, where the total power will be equal to three times the value measured on the meter.

Figure 1.35 Wattmeter connected to load

Measuring power in a three-phase, four-wire balanced load – one-meter method

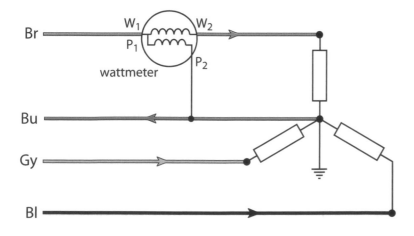

Figure 1.36 One-meter method

Measuring power in a three-phase balanced load – two-wattmeter method

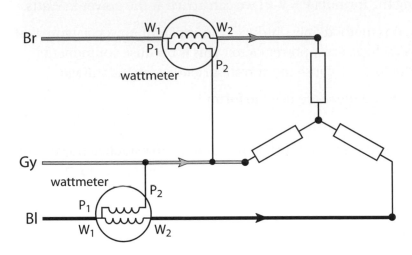

Figure 1.37 Two-meter method

In the two-wattmeter method, the total power is found by adding the two values together. At unity power factor the instruments will read the same and be half of the total load. For other power factors the instrument readings will be different, the difference in the reading could then be used to calculate the power factor.

Measuring power in an unbalanced three-phase circuit – three-wattmeter method

Here the total power will be the sum total of the readings on the three-wattmeters.

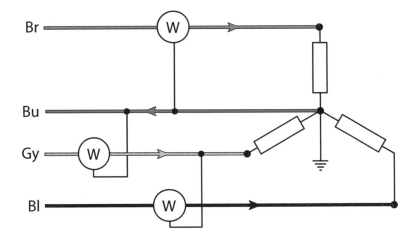

Figure 1.38   Three-phase circuit

This allows the power to be measured in a situation where the load is unbalanced, such as the three-phase supply to a large building where it is impossible to balance the load completely.

Measuring power factor

To measure power factor, there are a number of purpose made instruments available. All of these meters include both voltage and current measurement in circuits. Most designs are based on the clamp meter idea where the meter clamps around the current-carrying conductor. However, additional leads are then required to connect the meter across the supply. Many meters are also combined with the ability to measure all aspects of power, i.e. kW, kVA and kVAr.

The circuit diagram illustrates the connections of a power factor measuring instrument.

Figure 1.39   Circuit diagram with digital clamp meter

There is also an alternative method. The calculation to establish power factor is to divide true power by the apparent power (VA). As we know that a wattmeter will give us the true power, by adding an ammeter and a voltmeter to our circuit we can establish the apparent power (VA) and therefore establish the power factor. Figure 1.40 shows this arrangement.

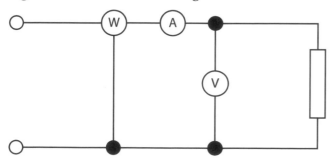

Figure 1.40   Circuit with wattmeter, ammeter and voltmeter connected

Knowledge check

1. A single-phase transformer with 2000 primary turns and 500 secondary turns is fed from a 230 volt a.c. supply. What will the secondary voltage and volts per turn in the secondary be?

2. What is the highest value in a sinusoidal wave form?

3. To calculate the average value of a sinusoidal voltage or current we must multiply the maximum value by what?

4. What is the correct formula for inductive reactance?

5. What is the correct formula for calculating impedance in a circuit containing resistance and inductance in series?

6. Power factor can be expressed as what?

7. Define reactive current.

8. Draw and label the power triangle.

9. On what principle to transformers work?

10. Power factor is the ratio of true power against what?

11. What is a c.t. used to measure?

12. From what is SELV derived?

Chapter 2

Earthing and protection

OVERVIEW

The purpose of earthing is to connect together all metalwork (other than that which is intended to carry current) to earth, so that dangerous potential differences cannot exist either between different metal parts, or between metal parts and earth. The definition given in BS 7671 for earthing is 'the act of connecting the exposed conductive parts of an installation to the main earthing terminal of that installation'.

By using the correct earthing procedures (and correct lightning protection measures), danger to life and the risk of fire to property can be greatly reduced.

This chapter will look at the following areas:

- **Purpose of earthing**
- **Earthing systems**
- **Lightning protection**
- **Electric shock**
- **Earthed equipotential bonding and automatic disconnection**
- **Residual current devices**
- **The earth-fault loop path**
- **Overcurrent protection**
- **Cable selection**

Purpose of earthing

By connecting to earth all metalwork not intended to carry current, a path is provided for leakage current which can be detected and interrupted by fuses, circuit breakers and residual current devices. Figure 2.1 illustrates the earth return path from the consumer's earth to the supply earth.

Figure 2.1 The earth-fault loop path

The connection at the consumer's earth can be by means of either an earth electrode at the building where the earth is required or may be in the form of a cable which runs back to the generator or transformer and is then connected to an earth point.

Because the transformer or generator at the point of supply always has an earth point, a circuit is formed when earth-fault currents are flowing. If these fault currents are large enough they will operate the protective device, thereby isolating the circuit. The star point of the secondary winding in a three-phase four-wire distribution transformer is connected to the earth to maintain the neutral at earth potential.

Results of an unearthed appliance

A person touching the appliance shown, which is live due to a fault, completes the earth circuit and receives an electric shock.

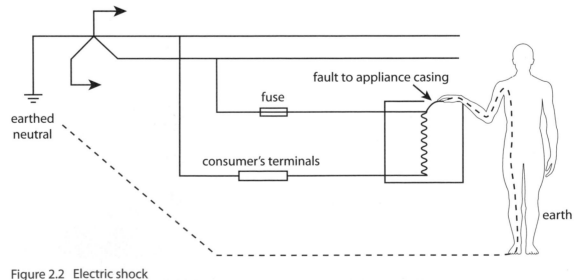

Figure 2.2 Electric shock

Results of a bad earth

A bad earth circuit, that is one with too large a resistance, can sometimes have more disastrous effects than having no earth at all. This is shown in the illustration, where the earth-fault circuit has a high resistance mainly due to a bad contact at point A.

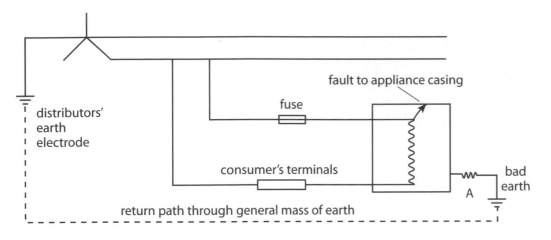

Figure 2.3   Bad earth path

The severity of shock will depend mainly upon the surroundings, the condition of the person receiving the shock and the type of supply. When the current starts to flow, the high resistance connection will heat up and this could be a fire hazard. Also, because the current flowing may not be high enough to blow the fuse or trip the circuit breaker, the appliance casing remains live.

Results of a good earth path

A good earth path, that is a low resistance one, will allow a high current to flow. This will cause the protective device to operate quickly, thereby isolating the circuit and giving protection against electric shock.

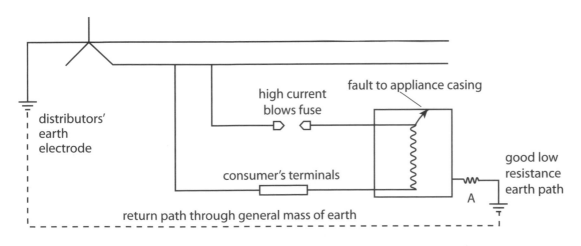

Figure 2.4   Good earth path

Earth-fault loop impedance

The path made or followed by the earth fault current is called the earth-fault loop or phase-earth loop. It is termed **impedance** because part of the circuit is the transformer or generator winding, which is inductive. This inductance, along with the resistance of the cables to and from the fault, makes up the impedance.

1 The circuit protective conductor (cpc) within the installation
2 The consumer's earthing terminal and earthing conductor
3 The earth return path, which can be either by means of an electrode or via the cable armouring
4 The path through the earthed neutral point of the transformer and the transformer winding (or generator winding)
5 The phase conductor.

Figure 2.5 Earth-fault loop impedance

Earthing systems

TT system

The letter T when used in this way of describing wiring systems comes from the Latin word 'terra', which means earth. So the first letter T in this instance means that the supply is connected directly to earth at the source. This could be the generator, or transformer at one or more points.

The second letter T means that the exposed metalwork of the installation is connected to the earth by a separate earth electrode. The only connection between these two points is the general mass of earth (soil etc.) as shown.

Figure 2.6 Earthing with customer's earth electrode and ground earth return path

When a fault to earth occurs on this system, the earth-fault current will flow around this circuit from the fault, through the earth and transformer windings, along the phase conductor and back to the fault position.

In this system the earth may have a very high value of impedance, for example 2300 ohms. From Ohm's Law we can see that the current flow would be:

$$I = \frac{V}{R} = \frac{230}{2300} = 0.1\ A$$

This current flow of 0.1 A, that is 100 mA, is sufficient to be fatal to people, so protection against even very small currents is vital to prevent danger from electric shock. Regulation 413-02-19 states that a residual current device (RCD) or an overcurrent protective device or both should protect a TT system, with the RCD being preferred. Figure 2.7 illustrates the earth connection.

Figure 2.7 Intake earthing arrangements on a TT system

TN-S system

- T means that the supply is connected directly to the earth at one or more points.

- N means the exposed metalwork of the installation is connected directly to the earthing point of the supply.

- S means a separate conductor is used throughout the system from the supply transformer all the way to the final circuit to provide the earth connection.

Figure 2.8 illustrates this system.

Figure 2.8 TN-S earthing system with metallic earth return path

This earth connection is usually through the sheath or armouring of the supply cable and then by a separate conductor within the installation. As a conductor is used throughout the whole system to provide a return path for the earth-fault current, the return path should have a low value of impedance. Figure 2.9 illustrates the intake earthing arrangements for a TN-S system.

final circuits

consumer unt (CCU)

energy meter

earthing conductor

cut out

earthing block

earth via armour of supply cable

Figure 2.9   TN-S earth connection

TN-C-S system

- T means the supply is connected directly to earth at one or more points.

- N means the exposed metalwork of the installation is connected directly to the earthing point of the supply.

- C means that for some part of the system (generally in the supply section) the functions of neutral conductor and earth conductor are combined in a single common conductor.

- S means that for some part of the system generally in the installation, the functions of neutral and earth are performed by separate conductors.

Figures 2.10 and 2.11 illustrate this system.

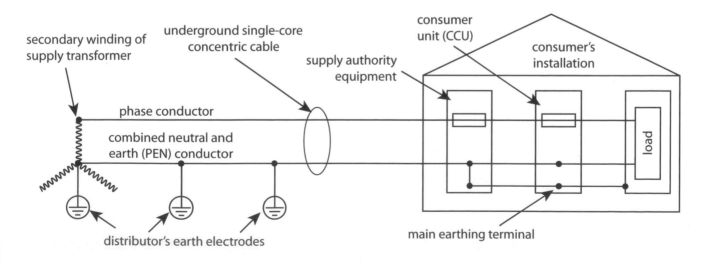

secondary winding of supply transformer

underground single-core concentric cable

consumer unit (CCU)

supply authority equipment

consumer's installation

phase conductor

combined neutral and earth (PEN) conductor

load

distributor's earth electrodes

main earthing terminal

Figure 2.10   TN-C-S, protective multiple earthing (PME) system

In a TN-C-S system, the supply uses a common conductor for both the neutral and the earth. This combined conductor is commonly known as the Protective Earthed Neutral (PEN) or also sometimes as the Combined Neutral & Earth (CNE) conductor.

In such a system the supply PEN is required to be earthed at several points. This type of system is also known as Protective Multiple Earthing (PME) and this will be discussed more fully in the next section.

However, commonly, as shown on page 49, this effectively means that the distribution system is TN-C and the consumer's installation is TN-S; this combination therefore giving us a TN-C-S system.

The intake earthing arrangements for a typical TN-C-S system are shown in Figure 2.11.

Figure 2.11 TN-C-S intake earthing arrangements

Protective Multiple Earthing (PME)

As shown in the previous section, this system of earthing is used on TN-C-S systems. PME is an extremely reliable system of earthing and is becoming the most commonly used distribution system in the UK today.

With the PME system, the neutral of the incoming supply is used as the earth point and all cpc's connect all metal work, in the installation to be protected, to the consumer's earth terminal.

Consequently, all phase to earth faults are converted into phase to neutral faults, which ensures that under fault conditions a heavier current will flow, thus operating protective devices rapidly.

However this increase in fault current may produce two hazards:

1. The increased fault current results in an enhanced fire risk during the time the protective device takes to operate.

2. If the neutral conductor ever rose to a dangerous potential relative to earth, then the resultant shock risk would extend to all the protected metalwork on every installation that is connected to this particular supply distribution network.

Because of these possible hazards, certain conditions are laid down before a PME system is used. These include the following:

- PME can only be installed by the supply company if the supply system and the installations it will feed meet certain requirements.

- The neutral conductor must be earthed at a number of points along its length. It is this action that gives rise to the name 'multiple' earthing.

- The neutral conductor must have no fuse or link etc. that can break the neutral path.

- Where PME conditions apply, the main equipotential bonding conductor shall be selected in accordance with the neutral conductor of the supply and Table 54H of BS 7671.

Other systems

TN-C system

In this system the neutral and protective functions are combined in a single conductor throughout the system. This system is relatively uncommon and its use is restricted for specific situations. There must be no metallic connection between this system and supply company equipment.

IT system

This system **must not** be connected to the supply company's system. It is a special system used in quarries, telephone exchanges and some industrial processes etc. The consumers must provide their own connection to earth.

FAQ

Q Why are there different supply systems?

A This is just recognition that in the past, before nationalisation of power supplies, there were independent suppliers of electricity, each with differing earthing arrangements to meet the requirements of their particular networks. TN-S was the most common but there were TN-C and TN-C-S systems as well. In the main all new-build projects are TN-C-S and as area supply systems are updated they get converted to TN-C-S. Remote and rural areas are supplied as TT systems as on overhead lines but more and more of these are being converted to TN-C-S. Whether there will ever be a time when all the supply systems are the same is uncertain.

Lightning protection

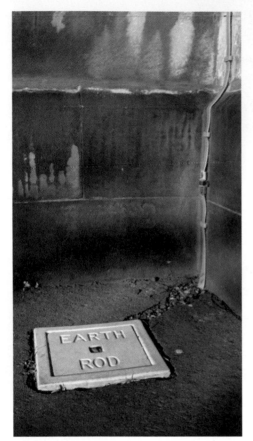

Lightning conductor connection

The protection of buildings and structures against lightning involves connecting various conductive parts to earth, hence its inclusion in this chapter. British Standard Code of Practice 326 (1965) and BS 6651 (1990) are the documents that cover the requirements for lightning protection and this type of installation is very specialised and is not something you would normally be involved with; however, an appreciation of the main requirements is useful.

The discharge current created when lightning strikes can be as high as 200,000 amps, which produces electrical, mechanical and thermal effects.

- The electrical effects happen when the lightning conductor's potential with respect to earth can become very high and 'flash-over' to other metalwork in the structure itself.

- The mechanical effects are caused by air pressure waves when the lightning first strikes or by the high discharge currents flowing to earth, which can cause large mechanical forces to be exerted on to the conductor, its fixings and supports.

- The thermal effects are caused by the rise in temperature of the lightning conductor. When the lightning strike happens a high value of current flows but only for a short time and the effect on the system is normally negligible.

The lightning protective system consists of a network of conductors (copper or aluminium) positioned along the roof and/or walls of a structure at specified distances, bonded to other parts of the structure such as radio and TV masts, and ultimately connected to a common point discharging to earth. The maximum resistance of this network of conductors should not exceed 10 ohms.

Electric shock

BS 7671 classifies electric shock into two categories – shock resulting from either:

- direct contact with the electrical supply
- indirect contact with the supply via exposed conductive parts or metalwork that have become live due to a fault.

Direct contact

Measures to be taken to prevent electric shock by direct contact include (Regulation 412):

(i)   protection by insulating live parts

(ii)   protection by barriers or enclosures

(iii) protection by obstacles, so preventing access

(iv) protection by placing out of reach

(v) IP codes/ratings are applied to electrical equipment enclosures to create an industry standard. The first number in the code relates to the protection against direct contact in terms of the degree of accessibility, e.g. IP2X means that there is no contact with a probe of more than 12.5 mm in diameter and less than 80 mm long, approximately the human finger. IPXXB is as above, but for probes up to a maximum of 80 mm.

The above measures are obvious but should not be disregarded. RCDs may be used as a supplementary means of shock prevention from direct contact, but only in addition to the measures listed in (i) to (v) above.

Indirect contact

In this situation, conductive parts, such as metalwork, have become live due to fault conditions. The potential voltage on this metalwork rises above that of earth and electric shock results when a person touches the metalwork.

Measures taken to prevent electric shock by indirect contact include (Section 413 in BS 7671):

(i) Earthed equipotential bonding and the automatic disconnection of the supply (EEBAD).

(ii) Use of class II equipment and/or equivalent insulation. Class II usually relates to portable equipment or factory-built equipment. Equipment to this class must not have enclosing metalwork earthed. Furthermore for a class II installation, close supervision is required to ensure the installation or its equipment is not changed from class II.

(iii) Non-conducting location, i.e. physical separation from exposed conductive parts, no earth connection.

(iv) Earth free local equipotential bonding. Every part is bonded and therefore 'no voltage difference – no shock'.

(v) Electrical separation. A transformer fed secondary is not connected to earth or an isolated generator supply is used.

Methods (iii) to (v) have a limited application, as close supervision is required. The design must be specified by a qualified electrical engineer. The basic limitation of these methods is the need to maintain their design condition. There should be no incorrect modifications, additions or deterioration.

Earthed equipotential bonding and automatic disconnection (EEBAD)

This is the electrical connecting of various exposed and extraneous conductive parts, within an installation, to earth to prevent a difference in potential occurring between them under fault conditions.

Exposed conductive parts

These are the conductive parts of an electrical system which can be touched and which are not normally live BUT they can become live under fault conditions.

Examples of exposed conductive parts are: metal casings of appliances such as heaters, kettles, ovens; wiring containment systems such as metal conduit, tray plate and trunking.

Extraneous conductive parts

These are conductive parts within a building which do not form part of the electrical system but can become live under fault conditions. Examples are copper water and gas pipes, air conditioning, boilers and air ducting systems.

Residual current devices

Residual current devices (RCDs) are a group of devices providing a modern approach to the enhancement of safety in electrical systems. They provide extra protection to people and livestock by reducing the risk of electric shock. Although RCDs operate on small currents, there are circumstances where the combination of operating current and high earth-fault loop impedance could result in the earthed metalwork rising to a dangerously high potential.

The regulations draw attention to the fact that if the product of operating current (A) and earth-fault loop impedance (Ω) exceeds 50V, the potential of the earthed metalwork will be more than 50V above earth potential and hence dangerous. This situation must not be allowed to occur.

Figure 2.12 Fault detector coil

Method of operation under fault conditions

The current taken by the load is fed through two equal and opposing coils wound onto a common iron core. Under normal safe working conditions, the current flowing in the live conductor into the load will be the same as that returning via the neutral conductor from the same load. When this happens, the phase and neutral conductors produce equal and opposing magnetic fluxes in the iron core contained within the RCD, resulting in no voltage being induced into the trip coil.

If more current flows in the phase side than the neutral side (or less returns via the neutral because the current has 'leaked' to earth within the installation) an out of balance flux will be produced which will induce a voltage into the fault detector coil. The fault detector coil then energises the trip coil and opens the double pole (DP) switch, due to the residual current produced by the induced voltage in the trip coil.

Functional testing of an RCD's operation should be carried out by operating the test button at regular intervals. Initial and periodic inspection and testing procedures additionally require the RCD to be tested using the appropriate instrument. RCDs are available for single and three phase applications.

Practical application of RCDs (Regulation 531-02):

1. RCDs must disconnect all live conductors at the same time (this includes the neutral).
2. All live conductors of the protected circuit are to be passed through the RCD's magnetic circuit (this includes the neutral).
3. An RCD should be selected so that any expected earth leakage current which may occur during normal operation of the connected loads (e.g. some computers) will be unlikely to cause unnecessary tripping of the device.
4. An RCD must be located outside stray magnetic fields.
5. RCDs are to be used to protect equipment being used outside the earthed equipotential zone, e.g. lawn mowers, hedge cutters, water pumps. This implies that a combination of Earthed Equipotential Bonding, Overcurrent Protection Devices, and Residual Current Devices in the design of an electrical installation should offer total electrical shock protection.

Sensitivity rated residual operating current	Level of protection	Applications
10 mA	Personnel	High risk areas: schools, colleges, workshops, laboratories. Areas where liquid spillage may occur
30 mA	Personnel	Domestic, commercial and industrial
100 mA	Personnel fire	Only limited personnel protection Excellent fire protection
300 mA	Fire	Commercial and industrial

Table 2.1 Overcurrent protection sensitivity

Residual Current Circuit Breaker with Overload Protection (RCBO)

An RCBO is essentially a marriage between a Miniature Circuit Breaker (MCB) and an RCD, as it comprises both of these components. Consequently, such a device offers protection against the effects of earth leakage, overload and short-circuit currents, while reducing the number of outgoing ways required within a distribution board.

Normally groups of circuits are protected by an RCD, and therefore all circuits would be interrupted under fault conditions, which can cause inconvenience. This device has the advantage of allowing earth-fault protection to be restricted to a single circuit, thus ensuring that only the circuit with the fault is interrupted.

A standard range of 10 A/30 mA, 16 A/30 mA, 20 A/30 mA, 32 A/30 mA and 40 A/30 mA are available from most manufacturers.

The earth-fault loop path

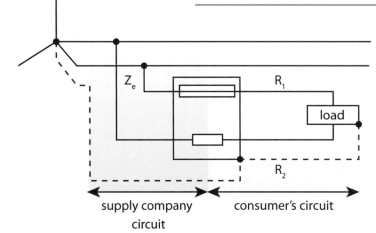

Figure 2.13 Earth-fault loop path

On a full earth-fault, the overcurrent protective device will operate, causing the automatic disconnection of the supply. On a low-level earth leakage fault, the overcurrent protective device will not detect the fault. With a faulty or disconnected earth-fault loop path, metalwork may become live under fault conditions.

An RCD will detect any imbalance between phase and neutral conductors, as any difference is assumed to be flowing to earth. This earth 'leakage' will cause the RCD to operate at its rated residual operating current.

Z_s is the sum of Z_e (the supply earth loop impedance) plus R_1 (the resistance of the circuit's phase conductor) and R_2 (the resistance of the circuit protective conductor), i.e.:

$$Z_s = Z_e + (R_1 + R_2)$$

$$\text{Fault current } I_f = \frac{U_o}{Z_s}$$

where U_o is the nominal supply voltage.

Note: This is also known as the Prospective Short Circuit Current (PSSC) I_{sc}. In situations where the value of Z_s is too high to achieve rapid disconnection on a fault then:

(i)   Cable conductor size may be increased to reduce the Z_s value.

(ii)   RCDs may be installed in combination with overcurrent protection devices and will provide a safe solution.

Earth-fault leakage currents will then be detected and the supply automatically disconnected.

The value of Z_s governs the flow of Earth-Fault Loop Current, the operation of earth-fault protective devices and their circuit disconnection times.

From Tables 41B1 and 41B2 and Appendix 3 Figures 3.3B, 3.5 & 3.6 of BS 7671 it can be seen that:

(i) A BS 88 40A fuse installed in a circuit of Z_s = 0.86 ohms will operate (at a fault current of 280A) in 0.4 seconds on a 230 volt system.

(ii) A BS EN 60898 40A Type C MCB installed in a circuit of Z_s = 0.6 ohms will operate (at a fault current of 400A) in 0.4 seconds on a 230 volt system.

(iii) A BS EN 60898 40A Type D MCB installed in a circuit of Z_s = 0.3 ohms will operate (at a fault current of 800A) in 0.4 seconds on a 230 volt system.

Regulations state that the protective device protecting a circuit must operate within a time of:

(i) 0.4 seconds for portable equipment supplied from socket outlet circuits
(ii) 5 seconds for fixed equipment
(iii) 0.2 seconds for construction sites and agricultural or horticultural premises.

Overcurrent protection

Protection devices

In a healthy electrical system, insulation between phase conductors and neutral conductors or between live conductors and earth is sound, with current flowing from the supply, through the load and returning via the neutral to its source.

Overcurrent protection is provided by means of a circuit breaker or fuse. These devices are designed to operate within specified limits, disconnecting the supply automatically in the event of an overload or fault current (short circuits or earth faults).

Earth-fault protection is also provided by means of a circuit breaker or fuse, which operates under earth-fault conditions. If a high value current flows these will safely disconnect the supply, but may not be relied upon to detect and disconnect the supply in the event of low values of earth leakage current.

RCDs are designed to detect and automatically disconnect the supply in the event of earth-faults. Earth-faults can exist, for example, when a live conductor touches an equipment case or if you cut through a cable when cutting the grass. They are potentially dangerous and must be eliminated in order to prevent hazards resulting from electric shock or abnormal temperature rises and fires in the electrical installation.

'Overcurrent protection device' applies to fuses and miniature circuit breakers and in this section we will be looking at the various means by which overcurrent protection devices provide protection to the cables and circuits from damage by too much current flowing. The appropriate device must operate within 0.4 or 5.0 seconds. The following areas will be looked at:

- overload and short circuit protection
- BS 3036 rewireable fuses
- BS 1361/1362 cartridge fuses
- BS 88 HBC fuses
- Type 'D' and Neozed fuses
- Miniature Circuit Breakers (MCBs)
- MCBs fusing factor
- fuse and circuit breaker operation
- prospective short circuit current
- discrimination.

Overload and short circuit protection

All circuit conductors must be sized correctly and must be of the correct current carrying capacity for the load that they have to carry. All circuits on an installation must be protected against **overcurrent**. This is a current exceeding the rated value or current carrying capacity of the cable conductor. Overcurrent conditions arise from either overloads or short circuits.

Overload current occurs in a circuit which is electrically sound, i.e. a circuit now carrying more current than it was designed to carry due to faulty equipment or too many pieces of equipment connected to the circuit. Overload is only a temporary fault condition and is easily corrected. Examples of how overload occurs include:

- several 13 A plugs connected to an adaptor and then plugged into one 13-amp socket outlet
- a stalled motor.

Such conditions cause the cable temperature to rise, resulting in an increased fire risk.

Fault current is the level of current flowing at a level up to and including the prospective short circuit current (the prospecive short circuit current being the maximum current that the supply transformer is capable of delivering).

The prospective short circuit current must also be taken into account when selecting the type of overcurrent device to be installed.

This fault occurs at a point between live conductors of negligible impedance or between a live conductor and an exposed conductive part. It is more serious than an overload current and will usually not be as easily corrected.

Examples of faults are: a wiring fault (live connected to neutral or earth by mistake); a nail or screw being driven through an energised cable, making contact with the live conductors and either neutral or earth conductors.

A fuse is the weak link in a circuit which will melt when too much current flows, thus protecting the circuit conductors from damage.

Calculation of the cable size therefore automatically involves the correct selection of protective devices.

BS 3036 rewireable fuses

Early rewireable fuses had a very low short circuit capacity and were very dangerous when operating under fault conditions because the fuse element melts and splashes the melted copper around and can cause fires. Later rewireable fuses incorporated asbestos to protect the fuse holder during the fusing period, thus reducing the risk of fire from scattering hot metal when rupturing.

BS 3036 rewirable fuse

The rewireable fuse consists of a fuse, holder, a fuse element and a fuse carrier. The holder and carrier are made of porcelain or bakelite. The circuits for which this type of fuse is designed have a colour code, marked on the fuse holder:

- 5 A white
- 15 A blue
- 20 A yellow
- 30 A red
- 45 A green.

This type of fuse was very popular in domestic installations although, with the exception of old installations, it is not normally used because it has some serious disadvantages.

Disadvantages of rewirable fuses	Advantages of rewirable fuses
• Easily abused when the wrong size of fuse wire is fitted **Fusing factor** of around 1.8 – 2.0, which means they aren't guaranteed to operate until up to twice the rated current is flowing. As a result cables protected by them must have a larger current carrying capacity • Precise conditions for operation cannot be easily predicted • Do not cope well with high short circuit currents • Fuse wire can deteriorate over time • Danger from hot scattering metal if the fuse carrier is inserted into the base when the circuit is faulty.	• Low initial cost • Can easily see when the fuse has blown • Low element replacement cost • No mechanical moving parts • Easy storage of spare fuse wire

Table 2.2   Advantages and disadvantages of rewireable fuses

Nominal current of fuse wire (A)	Nominal diameter of wire (mm)
3	0.15
5	0.2
10	0.35
20	0.5
20	0.6
25	0.75
30	0.85
45	1.25
60	1.53
80	1.8
100	2

Table 2.3   Size of tinned copper wire for use in semi-enclosed fuses

Table 2.3 indicates the size of tinned copper wire for use in semi-enclosed fuses.

BS 1361/1362 cartridge fuses

The cartridge fuse consists of a porcelain tube with metal end caps to which the element is attached. The tube is then filled with granulated silica. The BS 1362 fuse is generally found in domestic plug tops used with 13 A BS 1363 domestic socket outlets. There are two common fuse ratings available, the 3 A, which is for use with appliances up to 720 watts (radios, table lamps, electric blankets) and the 13 A fuse which is used for appliances rated over 720 watts (irons, kettles, fan heaters, electric fires, lawn mowers, toasters, refrigerators, washing machines and

BS 1361 and 1362 fuses

vacuum cleaners). There are, however, other sizes available (which are more difficult to come across). These are 1, 5, 7 and 10 amp sizes. The BS 1361 fuse is normally found in distribution boards and at main intake positions.

Disadvantages of cartridge fuses	Advantages of cartridge fuses
• They are more expensive to replace than rewireable fuses • They can be replaced with an incorrect size fuse (plug top type only) • The cartridge can be shorted out with wire or silver foil in extreme cases of bad practice • It is not possible to see if the fuse has blown • They require a stock of spare fuses to be kept	• They have no mechanical moving parts • The declared rating is accurate • The element does not weaken with age • They have a small physical size and no external arcing, which permits their use in plug tops and small fuse carriers • They have a low fusing factor around 1.6–1.8 • They are easy to replace

Table 2.4 Advantages and disadvantages of cartridge fuses

BS 88 high breaking capacity (HBC) fuses

A sectional view of a typical BS 88 HBC fuse

The HBC fuse is a sophisticated variation of the cartridge fuse and is normally found protecting motor circuits and industrial installations. It consists of a porcelain body filled with silica, a silver element and lug-type end caps. Another feature is the indicating bead, which shows when the fuse element has blown. It is a very fast acting fuse and can discriminate between a starting surge and an overload.

These types of fuses would be used when an abnormally high prospective short circuit current exists.

Disadvantages of BS 88 fuses	Advantages of BS 88 fuses
• They are very expensive to replace • Stocks of these spares are costly and take up space • Care must be taken when replacing them, to ensure that the replacement fuse has the same rating and also the same characteristics as the fuse being replaced	• They have no mechanical moving parts • The element does not weaken with age • Operation is very rapid under fault conditions • It is difficult to interchange the cartridge, since different ratings are made to different physical sizes

Table 2.5 Advantages and disadvantages of BS 88 fuses

Type 'D' and Neozed fuses

Neozed fuse

Both these fuses are manufactured in Germany and have been developed to European testing regulations where all European testing authorities have approved them. The Neozed is the successor to the 'D' type fuse. You may in the course of your work come into contact with either type of fuse.

Miniature Circuit Breakers (MCBs)

Due to improved design and performance, the modern MCB now forms an essential part of the majority of installations at the final distribution level. From about 1970 the benefits of current limiting technology have been incorporated into MCBs, thus providing the designer/user with predictable high performance overcurrent devices.

The circuit breaker is an automatic switch which opens when excess current flows in the circuit. The switch can be closed again when the current returns to normal because the device does not damage itself during normal operation. The contacts of a circuit breaker are closed against spring pressure, and held closed by a latch arrangement. A small movement of the latch will release the contacts, which will open quickly against the spring pressure to break the circuit.

The circuit breaker is so arranged that normal currents will not affect the latch, whereas excessive currents will move it to operate the breaker. There are two basic methods by which overcurrent can operate or 'trip' the latch.

Figure 2.14 Compression spring

Thermal tripping

The load current is passed through a small heater coil wrapped around a bi-metal strip inside the MCB housing; the heat created depends on the current it carries. This heater is designed to warm the bi-metal strip either directly (the current passes through the bi-metal strip which in effect is part of the electrical circuit) or indirectly (where a coil of current carrying conductor is wound around the bi-metal strip) and the excess current warms the bi-metal strip.

The bi-metal strip is made of two different metals, normally brass and steel (brass expanding more than steel). These two dissimilar metals are securely riveted or welded together along their length. The rate of expansion of the two metals is different so that when the strip is warmed, it will bend and will trip the latch.

The bi-metal strips are arranged so that normal currents will not heat the strip to tripping point.

However, if current increases beyond the rated value, the heater increases in temperature and thus the bi-metal strip is raised in temperature, trips the latch and opens the contacts.

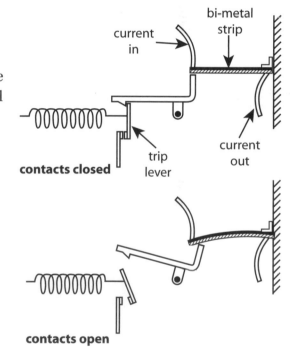

Figure 2.15 Thermal tripping action

Magnetic tripping

The principle used here is the force of attraction, which can be set up by the magnetic field of a coil carrying the load current. At normal current the magnetic field is not strong enough to attract the latch, but overload currents operate the latch and trip the main contacts.

Figure 2.16 A simple attraction type of magnetic trip

contacts

coil carrying current

magnetic plunger

oil filled dashpot

oil escape hole

Figure 2.17 Oil filled dashpot

The magnetic field is set up by a current in the flexible strip that attracts the strip to the iron. This releases the latch. This is often used in miniature circuit breakers and combined with a thermal trip.

The **oil dashpot** solenoid type is used on larger circuit breakers. The time lag is adjustable by varying the size of the oil-escape hole in the dashpot position. Current rating is adjustable by vertical movement of the plunger.

Combined tripping

As we have discussed, circuit breakers use two basic methods in their operation. There is always some time delay in the operation of a thermal trip, as the heat produced by the load current must be transferred to the bi-metal strip. Thermal tripping is therefore best suited to small overloads of comparatively long duration.

Magnetic trips are fast acting for large overloads or short circuits. The two methods are often combined to take advantage of the best characteristics of each.

They are then classified according to their instantaneous tripping current, i.e. the current at which they will operate within 100 ms. Types 1, 2, 3 and 4 are circuit breakers to BS 3871 and Types B, C and D are to BS EN 60898.

If you were to take for example a 20 A MCB of Types 1, 2, 3, and B, C, D and compare their instantaneous tripping currents, you would find that this will vary depending on the type of MCB chosen, and consequently this indicates that a 20 A MCB will trip instantaneously at different fault levels.

Guidance to the selection of circuit breakers to BS 3871 or BS EN 60898 is given in Table 7.2B of the *IEE On-Site Guide* as shown below, where I_n is the value of current that can be carried indefinitely by the device:

MCB type	Instantaneous trip current	Application
1	2.7 to 4 $\times I_n$	Domestic and commercial installations having little or no switching surge.
B	3 to 5 $\times I_n$	
2	4.0 to 7.0 $\times I_n$	General use in commercial/industrial installations, where the use of fluorescent lighting, small motors etc. can produce switching surges that would operate a Type 1 or B circuit-breaker. Type C or 3 may be necessary in highly inductive circuits such as banks of fluorescent lighting.
C	5 to 10 $\times I_n$	
3	7 to 10 $\times I_n$	
4	10 to 50 $\times I_n$	Suitable for transformers, X-ray machines, industrial welding equipment etc., where high in-rush currents may occur.
D	10 to 20 $\times I_n$	

Table 2.6 MCB selection

As can be seen from the table on page 62, Type 1 and B are the more sensitive devices. Please note that as part of the harmonisation of electrical standards, BS EN 60898 Types B, C and D will eventually replace the current BS 3871 types.

Disadvantages of MCBs	Advantages of MCBs
• They have mechanical moving parts • They are expensive • They must be regularly tested • Ambient temperature can change performance	• They have factory set operating characteristics which cannot be altered • They will maintain transient overloads and trip on sustained overloads • Easily identified when they have tripped • The supply can be quickly restored

Table 2.7 Advantages and disadvantages of MCBs

Fusing factor

It is evident that each of the protective devices discussed in the previous section provide different levels of protection, e.g. rewireable fuses are slower to operate and less accurate than MCBs. In order to classify these devices it is important to have some means of knowing their circuit breaking and 'fusing' performance. This is achieved for fuses by the use of a fusing factor.

$$\text{Fusing factor} = \frac{\text{Fusing current}}{\text{Current rating}}$$

This is the ratio of the fusing current, which is the minimum current that will cause the fuse to blow and the stated current rating of the fuse or MCB (which is the maximum current that the fuse can sustain without blowing). Fusing currents can be found in Appendix 3 of BS 7671. These tables are logarithmic and the scales increase by factors of ten, not uniformly as may be expected, and therefore the interpretation of these scales will require some practice. The rating of the fuse is the current it will carry continuously without deterioration.

Fusing factors for the above devices can generally be grouped as follows:
- BS 3036 1.8–2.0
- BS 88 1.25–1.7
- BS 1361 1.6–1.9
- MCBs up to 1.5.

The higher the fusing factor, the less accurate – and therefore less reliable – the device selected will be. You may, while looking at fuses, have noticed a number followed by the letters kA stamped onto the end cap of an HBC fuse or printed onto the body of a BS 1361 fuse. This is known as the breaking capacity of fuses and circuit breakers. When a short circuit occurs, the current may, for a fraction of a second, reach hundreds or even thousands of amperes.

The protective device must be able to break or make such a current without damage to its surroundings by arcing, overheating or the scattering of hot particles.

The breaking capacities of MCBs are indicated by an 'M' number e.g. M6. This means that the breaking capacity is 6 kA or 6,000 A. The breaking capacity will be related to the prospective short circuit current.

Fuse and circuit breaker operation

In this part the following symbols will be referred to:

I_z	this is the maximum current carrying capacity of a cable for continuous service, under the particular installation conditions concerned (Appendix 4 of BS 7671)
I_n	this is the nominal current or current setting of the device protecting the circuit against overcurrent (Appendix 3 of BS 7671)
I_2	this is the operating current, i.e. the fusing current or tripping current for the conventional operating time (Appendix 3 of BS 7671)
I_b	this is the design current of the circuit, i.e. the current intended to be carried by the circuit in normal service
I_{ef}	earth fault current
I_{sc}	short circuit current
t	time in seconds
k	material factor found from Tables 43A and 54B to 54F of BS 7671
s	conductor cross-sectional area in mm^2.

Consider a protective device (fuse or MCB) rated at 20 A. This value of current can be carried indefinitely by the device and is known as its nominal current setting, I_n. The value of current that will cause the device to operate, I_2, will be larger than I_n and will be dependent on the device's fusing factor.

This fusing factor figure, when multiplied by the nominal setting I_n, will give the value of operating current I_2.

For fuses to BS 1361, BS 88 and circuit breakers to BS 3871 and BS/EN 60898, the fusing factor has been approximated to 1.45. Therefore our 20 A device would operate when the current reached 29 A (1.45×20).

BS 7671 Regulation 433-02 requires co-ordination between conductor and protective device when an overload occurs such that:

(i) The nominal current or setting (I_n) of the protective device is not less than the design current (I_b) of the circuit, therefore $I_b \leq I_n$ (the symbol $\leq$ means less than or equal to). For example, if the circuit design current was 20 A then the protective device would need to have a rating of at least 20 A.

(ii) The nominal current or current setting (I_n) of the protective device does not exceed the lowest value of the current carrying capacities (I_z) of any of the conductors in the circuit. Therefore $I_n \leq I_z$.

The formulae can be combined so that $I_b \leq I_n \leq I_z$.

In simple terms this means that the design current of the circuit must be less than or equal to the protective device rating, which in turn must be less than or equal to the current carrying capacity of the cable.

(iii) The current (I_2) causing effective operation of the protective device does not exceed 1.45 times the lowest of the current carrying capacities (I_z) of any of the conductors of the circuit.

Therefore:

$I_2 \leq I_z \times 1.45$

We now have:

$I_2 = I_n \times 1.45$ and $I_2 \leq I_z \times 1.45$

We can now say:

$I_n \times 1.45 \leq I_z \times 1.45$

We can now remove the 1.45 from the equation, as it is common to both the left hand side (LHS) and the right hand side (RHS) of the equation.

Therefore, this leaves:

$I_n \leq I_z$.

This means that the current rating of the protective device must be less than or equal to the current carrying capacity of the cable.

This is only true for situations where BS 1361, BS 88 fuses and circuit breakers complying with BS 3871 and BS/EN 60898 are in use.

Now considering our 20 A device, if the cable is rated at 20 A then condition (ii) is satisfied. The fusing factor is 1.45; therefore condition (iii) is also satisfied, because:

$I_2 = I_n \times 1.45$ which is $20\,A \times 1.45 = 29\,A$

which is the same as $1.45 \times 20\,A$ (I_z) cable rating, because:

$1.45 \times 20\,A = 29\,A$ (rating of the cable).

Problems with BS 3036 rewireable fuses

This type of fuse may have a fusing factor as high as 2. Using the previous example we would see that:

$I_n \times 2 \leq I_z \times 1.45$ would not be true.

In order to comply with condition (iii) I_n should be less than or equal to $0.725\,I_z$. (This figure is derived from $1.45 \div 2 = 0.725$.)

For example if a cable is rated at 20 A, then I_n for a BS 3036 device should be less than or equal to:

$0.725 \times 20 = 14.5\,A$

As the fusing factor is 2, the operating current I_z would now be:

$2 \times 14.5 = 29\,A$

which conforms to condition (iii) if the fuse size was 14.5 A. However, if the design current required was the same as in the previous example and hence the fuse size (20 A) was the same, then this condition would not apply – the cable size (and hence its rating) would need to be increased. Be wary of using this type of fuse!

If all these requirements are met then this will ensure that the conductor insulation is undamaged when an overload occurs. These factors will be further explored in the cable selection section.

Prospective short circuit current (PSCC)

Regulation 434-01 states that 'protective devices shall be provided in a circuit to break any fault current flowing in conductors of that same circuit, before such current causes danger due to thermal or mechanical effects produced in those conductors or the associated connections'. This 'prospective fault current' must be taken into account when selecting the type of overcurrent device to be installed.

The effects of short circuit current are:

(i)   the thermal effect, which can cause melting of conductors/insulation, fire, alteration of the properties of materials etc.

(ii) the mechanical effects of large magnetic fields that can build up when short circuit currents are flowing, resulting in conductor distortion, breaking of supports/ insulators etc.

Rapid disconnection of the supply is essential to prevent this damage and is the worst case of short circuit fault. Possible causes of the occurrence of a short circuit fault are:

- contact between two poles of the supply due to incorrect connection

- equipment failure

- ingress of moisture

- accidental damage.

The causes of a fault are many and varied, e.g. it may result from the rupturing of a lamp filament or a JCB cutting an underground cable.

A short circuit is said to be a connection of two live conductors. 'Live conductors' means all those carrying current under normal conditions; **this includes the neutral conductor**.

A short circuit can occur between conductors connected to different phases, for example red to yellow or red to blue etc. or indeed any phase to neutral.

If a short circuit is two conductors touching, then it can be assumed that the resistance or impedance of that connection would be so low that it could be neglected.

Figure 2.18 Short circuit loop impedance

If a fault has negligible impedance, then the only restriction to the amount of current that will flow in the circuit is that of all the conductors. As conductors are of low resistance, then this total will be low and the current that flows can be very high.

Consequently, if for example a phase to neutral short circuit fault occurs within a final circuit as shown in Figure 2.18, the final circuit protective device should operate first.

Where:

- U_o = nominal supply voltage
- Z_e = loop impedance external to the installation
- R_1 = the resistance of the installation phase conductor to the fault point
- R_2 = the resistance of the installation return conductor to the fault point
- Z_s = total fault loop impedance,

$Z_s = Z_e + (R_1 + R_2)$.

Note the return can be via either the neutral or another phase depending on the supply arrangement.

Example 1

Calculate the prospective short circuit current (I_{sc}) flowing in the circuit shown below at:

(a) the customer consumer unit supply terminals

(b) the terminals of the load given the conditions shown.

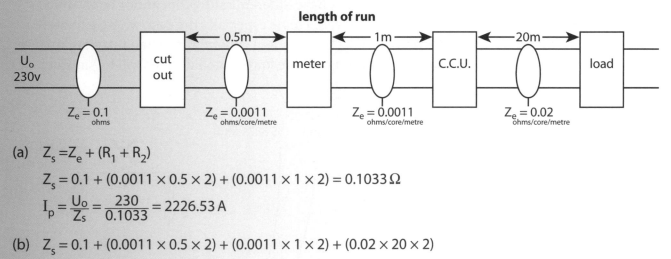

(a) $Z_s = Z_e + (R_1 + R_2)$

$Z_s = 0.1 + (0.0011 \times 0.5 \times 2) + (0.0011 \times 1 \times 2) = 0.1033\,\Omega$

$I_p = \dfrac{U_o}{Z_s} = \dfrac{230}{0.1033} = 2226.53\,A$

(b) $Z_s = 0.1 + (0.0011 \times 0.5 \times 2) + (0.0011 \times 1 \times 2) + (0.02 \times 20 \times 2)$

$= 0.1 + 0.0011 + 0.0022 + 0.8$

$= 0.9033\,\Omega$

$I_{sc} = 230 = 254.6\,A$

$Z_s = 0.9033\,\Omega$

Discrimination

In both large and small installations, there is usually a series of fuses and/or circuit breakers between the incoming supply and the electrical outlets. The relative rating of the protective devices used will decrease the nearer they are located to the current using equipment.

Ideally when a fault occurs, only the device nearest the fault operates, thus ensuring minimum disruption to other circuits not associated with the fault. **Discrimination** is said to have taken place when the smaller rated local device operates before the larger device.

Discrimination is generally only a problem when a system uses a mixture of devices. Obviously a particular type of fuse will discriminate against a similar type of fuse if it is of larger rating.

Discrimination is also known as the co-ordination between fuses.

Figure 2.19 relates to a 15 A semi-enclosed rewireable fuse serving a final sub-circuit, backed up by a 30 A MCB. When the fault current approaches 200 amps, the MCB will operate quicker than the fuse. Here discrimination has failed.

Simply because protective devices have different ratings, it cannot be assumed that discrimination will be achieved. This is especially the case where a mixture of different types of device is used. However, as a general rule when fuses are used in series in an installation a 2:1 ratio with the lower-rated devices will be satisfactory.

Figure 2.19 Discrimination

Cable selection

Cable selection can be defined as the 'rules' that you must follow when deciding which cable to choose for any installation.

This section will consider the following areas used in cable selection:

- external influences
- design current (I_b)
- rating of the protective device (I_n)
- reference methods
- correction factors
- application of correction factors
- voltage drop
- shock protection
- thermal constraints
- diversity.

External influences

Appendix 5 of BS 7671 lists the 'Classification of External Influences', where each condition of external influence is designated with a code that comprises a group of two capital letters and a number.

The first letter relates to the general category of the external influence (e.g. the Environment). The second letter then relates to the circumstances or nature of that external influence (e.g. the presence of Water). Finally, the number relates to the specific degree or level of that condition within each external influence (e.g. Splashes).

To explain this, let's look at the influences mentioned in the above paragraph and, using BS 7671 Table 5, convert them into a code.

General category of influence	Environment	A
Nature of the influence	Water	D
Level of that influence	Splashes	3

Consequently the designated code for these conditions would be: AD3

Working in the opposite direction and covering each of the three General Categories of external influence, here are three codes and their explanation:

AA4

A	Environment
AA	Environment – Ambient Temperature
AA4	Environment – Ambient Temperature – in the range –5°C to +40°C

BC1

B	Utilisation
BC	Utilisation – Contact with earth
BC1	Utilisation – Contact with earth – None

CA2

C	Construction of buildings
CA	Construction of buildings – Materials
CA1	Construction of buildings – Materials – Combustible

Once you have decided on the type of cable suitable for the environmental conditions, you must choose the size of conductor to be used. The first factor you need to consider is design current.

Design current (I_b)

You will have to calculate this. It is the normal resistive load current in the circuit. The following formulae apply to single and three phase supplies:

Single phase supplies:

$$U_o = 230\,V$$

$$I_b = \frac{Power}{U_o}$$

Where U_o is the phase voltage to earth (supply voltage).

Three phase supplies:

$$U_o = 400\,V$$

$$I_b = \frac{Power}{\sqrt{3} \times U_o}$$

In a.c. circuits, the effects of either highly inductive or highly capacitive loads can produce a poor power factor (PF) (inductive and capacitive loads will be explained later). You will have to allow for this. To find the design current, you may need to use the following equations:

Single phase circuits:

$$I_b = \frac{Power}{U_o \times PF}$$

Three phase circuits:

$$I_b = \frac{Power}{\sqrt{3} \times U_o \times PF}$$

Where PF is the power factor of the circuit concerned.

Rating of the protective device (I_n)

When you have worked out the design current of the circuit (I_b), you must next work out the current rating or setting (I_n) of the protective device.

Regulation 433-02-01, on page 57 in the *IEE Wiring Regulations* says that current rating (I_n) must be no less than the design current (I_b) of the circuit.

The reason for this is that the protective device must be able to pass enough current for the circuit to operate at full load, but without the protective device operating and disconnecting the circuit.

Protective devices come in standard values, for example 13 A for plug tops. You can find these values in the *IEE Wiring Regulations*, Tables 41B1, 41B2, 41D, 605B1 and 605B2, or in Appendix 3.

The table you use to select the value of your protective device will depend on the type of equipment or circuit to be supplied and the requirements for disconnection times.

Reference methods

Table 4A1 from the *IEE Wiring Regulations* lists all the methods that can be used to install a cable.

You need to decide at this stage in the cable selection process which method of installation to use. This will make sure that the correct cable column is chosen in the later stages of cable selection.

This choice is also important when you calculate correction factors for thermal insulation.

Correction factors

You, as the 'designer' of the installation, need to know the five correction factors, know where they are required and then apply these to the nominal rating of the protection (I_n).

Correction factor	Tables for correction factor values	Symbol
Ambient temperature	Tables 4C1 & 4C2	Ca
Grouping factors	Tables 4B1, 4B2 and 4B3	Cg
Thermal insulation	Regulation 523-04-01	Ci
BS 3036 fuse	0.725 and Table 4C2	Cr
Mineral insulated cable	0.9 Table 4J1A	N/A

Table 2.8 Tables for correction factor values

Ambient temperature

This is the temperature of the surroundings of the cable, often the temperature of the air in a room or building in which the cable is installed. When a cable carries current, it gives off heat. Therefore the hotter the surroundings of the cable, the more difficult it is for the cable to get rid of this heat. But if the surrounding temperature is low, then the heat given off could be easily let out and the cable could carry more current. Cables must give off this heat safely or they could be damaged and there is a risk of a fire. You can find the correction factor for **ambient temperature** in Tables 4C1 and 4C2 of the *IEE Wiring Regulations*.

These tables are based on an ambient temperature of 30°C. This means that any cables installed in an ambient temperature above this will need the correction factor applying to them. This is because the cable will not be able to get rid of the heat it gives off safely when carrying current.

Note that an ambient temperature below 30°C (e.g. 25°C) will give a correction factor that will allow your choice of cable to be improved. Because a lower ambient temperature will allow the cable to 'give off' heat more easily and safely therefore you could reduce the size of cable.

When a cable runs through areas having different ambient temperatures, correction factors should be applied to the highest temperature only. You should also note that Table 4C2 is only to be used where the protection device is a BS 3036 rewireable fuse as this is considered an unreliable protective device, because the fuse element is open and exposed.

Table 4C1 is for a system that has a protective device other than a BS 3036 fuse, for example a cartridge fuse to BS 88, BS 1361, BS 1362, or MCBs.

Grouping factors

If a number of cables are run together and touching each other, they will all produce heat when they are carrying current. The effect of this is that they are less able to cool down. So when installing cables we should keep them separated to reduce overheating.

If this is impossible, then you must apply a grouping factor to the cables. Grouping factors are in Tables 4B1, 4B2 and 4B3. You will see from the notes with these tables that the method of installation can affect the correction factor to be used:

- If each cable is separated by a clearance to the next surface of at least one cable diameter (De) they are classified as 'spaced'. You must take care in choosing the correction factor.

- Where the horizontal clearance to the next cable is more than 2De then no correction factor is needed.

It is important that you apply these correction factors to the number of circuits or multicore cables that are grouped and not the number of conductors. You will find that you are normally using Table 4B1.

Thermal insulation

Thermal insulation has the effect of wrapping a cable in a fur coat on a hot summer's day. The heat produced when the cable carries current cannot escape. There are two ways that thermal insulation is relevant in electrical installation:

- Situations where the cable is in contact with thermal insulation on one side of the cable only. In this case, you must choose 'method 4' or 'method 6' for a cable type covered by Table 4D5A of BS 7671 from the table of reference methods.

- Situations where the cable is totally enclosed in thermal insulation for any length from 50 mm up to 500 mm. The cable must have a correction factor applied as stated in Regulation 523-04-01 Table 52A (see Table 2.9). If more than 500 mm of the cable is to be totally enclosed, then the correction factor to be applied must be 0.5. The cable rating is then effectively halved.

Length in insulation (mm)	De-rating factor
50	0.89
100	0.81
200	0.68
400	0.55

Table 2.9 Cable rating

BS 3036 fuse

You could say that the only device that does not disconnect in sufficient time is the BS 3036 semi-enclosed rewireable fuse.

If this particular protective device is used, then we must always use a factor of 0.725 when calculating current carrying capacity.

Mineral insulated cable

Table 4J1A (BS 7671) states that, for bare cables (i.e. no PVC outer covering) exposed to touch, the tabulated values should be multiplied by 0.9.

Table 4J2A states that no correction factor for grouping need be applied.

Application of correction factors

You will have seen that nearly all of these factors are less than 1. We also know that the cable must be able to carry current of an equal or greater value than the rating of the fuse that protects the cable.

Therefore, the purpose of applying these factors is to make sure that the cable's conductor is large enough to carry the current without too much heat being generated.

If the cable's ability to give off heat is reduced by external conditions, then the only solution is to increase the size of the cable.

The use and application of the correction factors will require you to increase the physical cross sectional area of the cable, which in turn increases the cable's capacity to carry current and give off heat.

When the right correction factors have been applied, we then know the effective current carrying capacity of the conductor (I_z).

We then need to apply the result from the formula to the correct cable table to find the correct cable size.

This value of current tabulated in Appendix 4 of BS 7671 is given the symbol I_t which must be greater than I_z. Hence:

$$I_t \geq I_z \geq \frac{I_n}{C_a \times C_g \times C_i \times C_r}$$

(if they apply)

Now you can use the value of I_z, together with the correct cable table from BS 7671 Table 4D1A to 4L4B, to find the size of cable that you need to use, given these correction factors.

There are some examples at the end for you to practise working with correction factors. However, this is not the end of cable selection and you must check for voltage drop, shock protection and thermal constraints. All these subjects will be covered in the next section.

Example 1

A single-phase circuit has a design current of 30A. It is to be wired in flat two core 70°C PVC insulated and sheathed cable with cpc to BS 6004. It will have copper conductors and be grouped with four other similar cables, all clipped direct. If the ambient temperature is 30°C and the circuit is to be protected by a BS 3036 fuse, what should be the nominal current rating of the fuse and the minimum csa of the cable conductor?

In answering these types of questions, it is a good idea to construct a table of information. This will help you understand and retain the process of cable selection. It also makes cable selection easier to understand.

Installation Method No. is 1 (Clipped direct), therefore Reference Method = 1

From Table 4B1, as grouped, $C_g = 0.6$ (there being 5 circuits)

From Table 4C2, as 70° PVC, $C_a = 1.0$ (as 30° ambient temperature)

Design Current $(I_b) = 30$ A

As the protective device rating (I_n) must be $\geq I_b$, we can take $I_n = 30$ A

$$I_z = \frac{30}{0.6 \times 1.0 \times 0.725} = 69 \text{ A}$$

$I_t \geq I_z$, and from Table 4D5A column 4, it is found that the minimum conductor csa that can be used is 16 mm^2 cable with a tabulated value (I_t) of 85 A.

Over the following pages we will develop this example further to include compliance with volt drop, shock protection and thermal protection.

Voltage drop

Regulation 525-01-02 states 'the **voltage drop** between the origin of the installation (usually the supply terminals) and a socket outlet or the terminals of the current using equipment should not exceed 4 per cent of the supply voltage'.

This means that with a supply voltage of 230 volts, a voltage drop of 9.2 volts (which is four per cent of 230) is the maximum that can be allowed.

The voltage drops either because the resistance of the conductor becomes greater as the length of the cable increases or the cross-sectional area of the cable is reduced. This means that on long cable runs, the cable cross-sectional area may have to be increased, reducing the resistance allowing current to 'flow' more easily and reducing the voltage drop across the circuit.

You can calculate the maximum permissible voltage drop allowed like this:

1. For single phase 230 volt systems:

$$4\% = \frac{230 \times 4}{100} = 9.2 \text{ volts}$$

2. For three phase 400 volt systems:

$$4\% = \frac{400 \times 4}{100} = 16 \text{ volts}$$

Therefore, when calculating with cable runs, we are allowed to have a system that gives as little as 220.8 volts on single phase and 384 volts on three phase. It is better to keep the voltage drops as low as possible, because low voltages can reduce the efficiency of the equipment being supplied. Lamps will not be as bright and heaters may not give off full heat. The values for cable voltage drop are given in the accompanying tables of current carrying capacity in Appendix 4 of the *IEE Wiring Regulations*. The values are given in milli volts per ampere per metre(mV/A/m).

You should use the formula below to calculate the actual voltage drop.

$$\text{Voltage drop (VD)} = \frac{\text{mV/A/m} \times I_b \times L}{1000}$$

Where mV/A/m   is the value given in the Regulation Tables.

I_b is the circuit's design current.

L   is the length of cable used in the circuit measured in meters.

Note: Alternatively in some areas of the country people prefer the following equation for working out voltage drop.

$$\text{Voltage drop (VD)} = \text{mV/A/m} \times I_b \text{ x } L \times 10^{-3}$$

Example 2

This is the same question that was used in Cable Selection Example 1 but this time you are also going to work out the voltage drop on this cable.

A single-phase circuit has a design current of 30A and is to be wired in flat two core 70°C PVC insulated and sheathed cable with cpc to BS 6004. It will have copper conductors and be grouped with four other similar cables, all clipped direct. If the ambient temperature is 30°C and the circuit is to be protected by a BS 3036 fuse, what should be the nominal current rating of the fuse and the minimum csa of the cable conductor? Assume the cable run is 27 m.

Installation Method No. is 1 (Clipped direct), therefore Reference Method = 1

From Table 4B1, as grouped, $C_g = 0.6$ (there being 5 circuits)

From Table 4C2, as 70° PVC, $C_a = 1.0$ (as 30° ambient temperature)

Design Current (I_b) = 30 A

As the protective device rating (I_n) must be $\geq I_b$, we can take $I_n = 30$ A

$$I_z = \frac{30}{0.6 \times 1.0 \times 0.725} = 69 \text{ A}$$

$I_t \geq I_z$, and from Table 4D5A column 4, it is found that the minimum conductor csa that can be used is 16mm² cable with a tabulated value (I_t) of 85A.

From Table 4D5A the voltage drop (VD) is 2.8 mV/A/m

$$\text{Therefore VD} = \frac{2.8 \times 30 \ (I_b) \times 27 \ (\text{length})}{1000} = 2.27 \text{ volts}$$

Actual VD must not exceed 4% of the supply voltage. 4% of 230V = 9.2V. Therefore, for this example the volt drop of 2.27V is acceptable.

Shock protection

The next stage in cable selection is to check that the protective device (fuse or MCB) will work inside the specified time limits if there is a fault. This gives the cable and circuit protection from fault current.

The speed of operation of the protective device is extremely important. It will depend on the size of the fault current. This in turn will depend on the impedance of the earth fault loop path.

The value of earth fault loop impedance (Z_s) should not be larger than the values given in Tables 41B1, 41B2, 41D, 604B1, 604B2, 605B1 and 605B2.

Tables 2.12 gives the maximum allowed value of earth fault loop impedance for a given protective device for a given disconnection time.

Situation	Table	Max disconnection time
Construction sites	604 B1	0.2 sec
	604 B2	0.2 sec
Agricultural and horticultural premises	605 B1	0.2 sec
	605 B2	0.2 sec
Socket outlets	41 B1	0.4 sec
	41 B2	0.4 sec
Fixed equipment	41 B3	5 secs

Table 2.10   Maximum allowed value of earth loop impedance

You need to do a calculation to check that the circuit protective device will operate within the required time.

You do this by checking that the actual Z_s is lower than the maximum Z_s given in the relevant table, for the protective device you have chosen.

Maximum Z_s can also be found from *IEE Wiring Regulations* or manufacturers' data. Use the following formula to calculate the actual Z_s:

$$\text{Actual } Z_s = \frac{Z_e + (\{R_1 + R_2\} \times \text{multiplier (mf)} \times \text{length})}{1000}$$

Where:

- Z_e = the external impedance on the supply authority's side of the earth fault loop. You can get this value from the supply authority. Typical maximum values are: TN-C-S (PME) system 0.35 ohms, TN-S (cable sheath) 0.8 ohms, TT system 21 ohms.

- $R_1 + R_2$ = the resistance of the phase conductor plus the cpc resistance. You can find values of resistance/metre for $R_1 + R_2$, for various combinations of phase and cpc conductors up to and including 50 mm in Table 9A of the *IEE On-Site Guide* or Table G1–G2 in the Amicus Guide.

- Tables 9B and 9C of the *IEE On-Site Guide* give you multipliers (mf) to apply to the values given in Table 9A to calculate the resistance under fault conditions. If the conductor temperature rises, resistance in the conductors will increase. Tables 9B and 9C are based on the type of insulation used and whether the cpc is, or is not: incorporated in a cable; bunched with the cables; bare and in contact with the cable covering (54B); a core in a cable; bunched with cables (54C).

This information, together with the length of the circuit, can now be applied to the formula. Then you can check to see that the actual Z_s is less than the maximum Z_s given in the appropriate tables.

When you have done this, you can prove that the protective device will disconnect the circuit in the time that is specified in the appropriate table. In other words, if the actual Z_s is less than the maximum Z_s we have compliance for shock protection.

Example 3

This is the same question that was used in Cable Selection Examples 1 and 2. This time you are also required to work out the requirements for shock protection. Assume the cable supplies a domestic cooker and the Z_e is 0.8.

A single-phase circuit has a design current of 30A and is to be wired in flat two core 70°C PVC insulated and sheathed cable with cpc to BS 6004. It will have copper conductors and be grouped with four other similar cables, all clipped direct. If the ambient temperature is 30°C and the circuit is to be protected by a BS 3036 fuse, what should be the nominal current rating of the fuse and the minimum csa of the cable conductor? Assume the cable run is 27m.

Installation Method No. is 1 (Clipped direct), therefore Reference Method = 1

From Table 4B1, as grouped, $C_g = 0.6$ (there being 5 circuits)

From Table 4C2, as 70° PVC, $C_a = 1.0$ (as 30° ambient temperature)

Design Current $(I_b) = 30$ A

As the protective device rating (I_n) must be $\geq I_b$, we can take $I_n = 30$ A

$$I_z = \frac{30}{0.6 \times 1.0 \times 0.725} = 69\,A$$

$I_t \geq I_z$, and from Table 4D5A column 4, it is found that the minimum conductor csa that can be used is 16 mm^2 cable with a tabulated value (I_t) of 85 A.

From Table 4D5A the voltage drop (VD) is 2.8 mV/A/m

$$\text{Therefore VD} = \frac{2.8 \times 30\,(I_b) \times 27\,(\text{length})}{1000} = 2.27 \text{ volts}$$

Actual VD must not exceed 4% of the supply voltage. 4% of 230V = 9.2V. Therefore, for this example the volt drop of 2.27V is acceptable.

From Table 41D, the maximum Z_s for a 30 A BS 3036 fuse is 2.76W.

$$\text{EFLI } (Z_s) = Z_e + \frac{([R_1 + R_2] \times \text{any multipliers} \times \text{length})}{1000}$$

$$Z_e = 0.8$$

From Table 9A of the *IEE On-Site Guide*, 16 mm^2 phase + 6 mm^2 cpc

$$R_1 + R_2 = 4.23 \text{ mW/m}$$

From Table 9C of the *IEE On-Site Guide*, the multiplier for a grouped cable is 1.20

$$\text{Therefore, } Z_s = 0.8 + \frac{([4.23] \times 1.2 \times 27)}{1000} \text{ giving } Z_s = 0.8 + \frac{(137.05)}{1000} = 0.937\,W$$

As the actual value of Z_s is less than the permitted value (2.76W), our cable of 16 mm^2 is acceptable.

Thermal constraints

Now that you have chosen the type and size of cable to suit the conditions of the installation, we must look at 'thermal constraints'. This is a check to make sure that the size of the cpc, 'the earth conductor', complies with the *IEE Wiring Regulations*.

If there is a fault on the circuit, which could be a short circuit, or earth fault, a fault current of hundreds or thousands of amperes could flow. Imagine that this is a 1 mm^2 or 2.5 mm^2 cable; if this large amount of current was allowed to flow for a short period of time, i.e. a few seconds, the cable would melt and a fire would start.

We need to check that the cpc will be large enough to be able to carry this fault current without causing any heat/fire damage. The formula that is used to check this situation is the **adiabatic equation**. The cpc will only need to carry the fault current for a short period of time, until the protective device operates.

Regulation 543-01-01 states that 'The cross-sectional area of every protective conductor shall be calculated in accordance with Regulation 543-01-03 (adiabatic equation) or selected in accordance with Regulation 543-01-04 (Table 54G)'. You will see that Regulation 543-01-04 asks that reference should be made to Table 54G. This table shows that, for cables 16 mm^2 and below with the cpc made from the same material as the phase conductor, the cpc should be the same size as the phase conductor. A phase conductor between 16 mm^2 and 35 mm^2 requires a cpc to be 16 mm^2. A phase conductor above 35 mm^2 requires a cpc to be at least half the cross-sectional area.

Multicore cables have cpcs smaller than their respective phase conductors, except for 1 mm^2, which has the same-sized cpc. Regulation 543-01-02 of BS 7671, gives two options, calculation or selection. If we were to apply option (ii) of this Regulation then selection would make these cables contravene the Regulations. Clearly, it is not intended that composite cables should have their cpcs increased in accordance with the table and therefore calculation by the adiabatic equation required in option (i) of the same Regulation should be applied.

On the job: Intermittent fault

You have been asked by your employer to investigate an 'earthing' problem that has been reported by a customer. The customer concerned has just moved into a large 80-year-old house that will allow him to work from home as a computer software designer. The house has large gardens, a separate garage, three outside security lights, a gas-fired central-heating system and the customer will be working from one of the bedrooms on his computers.

The customer has reported that he is getting intermittent RCD trips on the main consumer unit. The only things the customer has noticed is that the trips seem to occur only at night, and only when it has been raining. Rain during the day does not cause the same problems. There is also a circuit out to the garage that has its own RCD and this does not trip.

1   What do you think could be causing the problem?

(A suggested answer can be found on page 314)

The adiabatic equation referred to in the introduction enables a designer to check the suitability of the cpc in a composite cable. If the cable does not incorporate a cpc, a cpc installed as a separate conductor may also be checked.

The equation is as follows:

$$S = \frac{\sqrt{I_{EF} \times t}}{k}$$

Where:

S = the cross-sectional area of the cpc in mm^2.

I_{EF} = the value of the fault current I for earth fault current (this is sometimes referred to as I_a but for the purpose of these notes we shall use I_{EF}).

t = the operating time of the disconnecting device in seconds.

k = a factor depending on the conductor and its insulating material.

In order to apply the adiabatic equation, we first need to calculate the value of I (fault current) from the following equation:

$$I = \frac{U_o}{Z_s}$$

Where:

U_o is the nominal supply voltage.

Z_s is the earth fault loop impedance.

If you are using method (i) from Reg 543-01-02 and applying the adiabatic equation, you must find out the time/current characteristics of the protective device. A selection of time/current characteristics for standard overcurrent protective devices is given in Appendix 3 of *IEE Wiring Regulations*. You can get the time (t) for disconnection to the corresponding earth fault current from these graphs.

If you look at the time/current curve you will find that the scales on both the time (seconds) scale and the prospective current (amperes) scale are logarithmic and the value of each subdivision depends on the major division boundaries into which it falls.

For example, on the current scale, all the subdivisions between 10 and 100 are in quantities of 10, while the subdivisions between 100 and 1000 are in quantities of 100 and so on. This also occurs with the time scale, subdivisions between 0.01 and 0.1 being in hundredths and the subdivisions between 0.1 and 1 being in tenths, etc.

As an example, if you look at the graph shown in Figure 2.20 you will see that for a BS 88 fuse with a rating of 32 A, a fault current of 200 A will cause the fuse to clear the fault in 0.6 seconds.

In addition to this graph, the IEE has produced a small table at the top right-hand corner showing some of the more common sizes of protective devices and the fault currents for a given disconnection time.

Next, you need to select the k factor using Tables 54B to 54F. The values in these tables are based on the initial and final temperatures shown in each table. This is where you may need to refer to the cable's operating temperature shown in the cable tables.

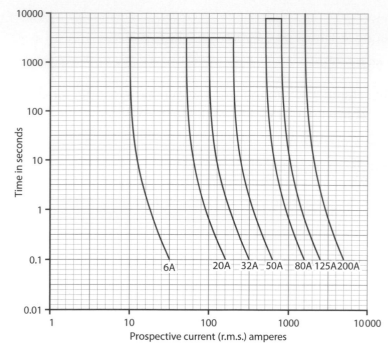

Time/Current characteristics for fuses to BS 88: Part 2 and Part 6				
Fuse rating	**Current for time**			
	0.1 sec	**0.2 sec**	**0.4 sec**	**5 secs**
6A	36A	31A	27A	17A
20A	175A	150A	130A	79A
32A	320A	250A	220A	125A
50A	640A	450A	380A	220A
80A	1100A	890A	740A	400A
125A	1800A	1500A	1300A	680A
200A	3000A	2500A	2200A	1200A

Figure 2.20 Time/current characteristics graph

Now you must substitute the values for I, t and k into the adiabatic equation. This will give you the minimum cross-sectional area for the cpc. If your calculation produces a non-standard size, you must use the next largest standard size.

From a designer's point of view, it is advantageous to use the calculation method as this may lead to savings in the size of cpc.

Now try the following example to give you a complete understanding of cable selection. The example is similar to those used in Cable Selection Examples 1, 2 and 3, but this time you also need to complete calculations for thermal constraints.

Example 4

A single-phase circuit feeding a domestic cooker has a design current of 30 A and a Z_e value of 0.8. It is to be wired in flat two core 70°C PVC insulated and sheathed cable with cpc to BS 6004. It will have copper conductors and be grouped with four other similar cables, all clipped direct. If the ambient temperature is 30°C and the circuit is to be protected by a BS 3036 fuse, what should be the nominal current rating of the fuse and the minimum csa of the cable conductor? The cable run is 27 m in length.

Installation Method No. is 1 (Clipped direct), therefore Reference Method = 1

From Table 4B1, as grouped, C_g = 0.6 (there being 5 circuits)

From Table 4C2, as 70° PVC, C_a = 1.0 (as 30° ambient temperature)

Design Current (I_b) = 30 A

As the protective device rating (I_n) must be $\geq I_b$, we can take $I_n = 30\,A$

$$I_z = \frac{30}{0.6 \times 1.0 \times 0.725} = 69\,A$$

$I_t \geq I_z$, and from Table 4D5A column 4, it is found that the minimum conductor csa that can be used is 16 mm² cable with a tabulated value (I_t) of 85 A.

From Table 4D5A the voltage drop (VD) is 2.8 mV/A/m

$$\text{Therefore VD} = \frac{2.8 \times 30\,(I_b) \times 27\,(\text{length})}{1000} = 2.27 \text{ volts}$$

Actual VD must not exceed 4% of the supply voltage. 4% of 230V = 9.2V. Therefore, for this example the volt drop of 2.27V is acceptable.

From Table 41D, the maximum Z_s for a 30 A BS 3036 fuse is 2.76 W.

$$\text{EFLI}\,(Z_s) = Z_e + \frac{([R_1 + R_2] \times \text{any multipliers} \times \text{length})}{1000}$$

$$Z_e = 0.8$$

From Table 9A of the *IEE On-Site Guide*, 16 mm² phase + 6 mm² cpc

$$R_1 + R_2 = 4.23\,\text{mW/m}$$

From Table 9C of the *IEE On-Site Guide*, the multiplier for a grouped cable is 1.20

$$\text{Therefore, } Z_s = 0.8 + \frac{([4.23] \times 1.2 \times 27)}{1000} \text{ giving } Z_s = 0.8 + \frac{(137.05)}{1000} = 0.937\,W$$

As the actual value of Z_s is less than the permitted value (2.76W), our cable of 16 mm² is acceptable.

To apply the adiabatic equation, $I = \dfrac{U_o}{Z_s} = \dfrac{230}{0.937} = 245.46\,A$

From Fig. 3.2A of Appendix 3 of BS 7671, t = 0.1 second. From Table 54C for 70° PVC, k = 115

$$S = \frac{\sqrt{I^2 \times t}}{k} = \frac{\sqrt{245.46^2 \times 0.1}}{115} = \frac{77.62}{115} = 0.68\,\text{mm}^2$$

Therefore our 6 mm² cpc is acceptable.

Diversity

In this section we will look at Maximum Demand and Diversity as required in accordance with the *IEE On-Site Guide*. The current demand for a final circuit is determined by adding up the current demands of all points of utilisation and equipment in the circuit and, where appropriate, making an allowance for diversity. Typical demands to be used for this summation are given in Appendix 1, Table 1A of the *IEE On-Site Guide*.

In most cases main or sub-main cables will supply a number and/or variety of final circuits. Use of the various loads must now be considered, otherwise if all the loads are totalled, a larger cable than necessary will be selected at considerable extra cost. Therefore a method of assessing the load must be used.

Appendix 1, Table 1B, of the *IEE On-Site Guide* contains such a method, which allows diversity to be applied depending upon the type of load and installation premises. The individual circuit/load figures are added together to determine the total 'Assumed Current Demand' for the installation. This value can then be used as the starting point to determine the rating of a suitable protective device and the size of cable, considering any influencing factors in a similar manner to that applied to final circuits.

Please also remember that the calculation of maximum demand is not an exact science and a suitably qualified electrical engineer may use other methods of calculating maximum demand. This subject is also discussed under 'Load balancing' in Book 1.

Example

A 230 volt domestic installation consists of the following loads:
- 15 x filament lighting points
- 6 x fluorescent lighting points, each rated at 40 watts
- 4 x fluorescent lighting points each rated at 85 watts
- 3 x ring final circuits supplying 13 A socket outlets
- 1 x A3 radial circuit supplying 13 A socket outlets for adjoining garage
- 1 x 3 kW immersion heater with thermostatic control
- 1 x 13.6 kW cooker with a 13 A socket outlet incorporated in the control unit.

Determine the maximum current demand for determining the size of the sub-main cable required to feed this domestic installation? The circuit protection is by the use of BS 1361 fuses.

Answer

Lighting

Tungsten light points (See Table 1A of the *IEE On-Site Guide*)
15 x 100W minimum	1500

Fluorescent light points (See Note 2 below Table 1A of the *IEE On-Site Guide*)
6 x 40W with multiplier of 1.8 (40 x 1.8 = 72)	432
4 x 85W with multiplier of 1.8 (85 x 1.8 = 153)	612
Total	2544 W

Using Item 1 in Table 1B of the *IEE On-Site Guide*, we can apply diversity as being 66% of the total current demand. Therefore:

66% of 2544 = 1679 W   and since $I = \dfrac{P}{V}$ this gives us $\dfrac{1679}{230} =$ 7.3 A

Power (See Item 9 in Table 1B of the *IEE On-Site Guide*)

3 Ring final circuits
1 x ring at 100% rating (30 A)	30 A
2 x ring at 40% rating (40% of 30 A)	24 A

20 A A3 radial circuit

1 × radial at 40% rating (40% of 20 A)   8 A

3 kW Immersion Heater (See Item 6 in Table 1B of the *IEE On-Site Guide*)

3 kW heater with no diversity $I = \dfrac{P}{V} = \dfrac{3000}{230}$ which gives us   13 A

13.6 kW Cooker with socket outlet (See Table 1A and Item 3 in Table 1B of the *IEE On-Site Guide*)

The first 10 A, plus 30% of the remainder of the overall rated current, plus 5 A for the socket.

$I = \dfrac{P}{V}$ giving $\dfrac{136000}{230}$ which gives us a total rated current of 59 A

59 A – 10 A = 49 A   and therefore 30% of 49 A = 14.7 A

The allowable total cooker rating is therefore   10 A + 14.7 A + 5 A =   29.7 A

Our total assumed current demand is therefore :

7.3 + 30 + 24 + 8 + 13 + 29.7   giving a total of   **112 A**

Knowledge check

1. What could be the result of a bad earth in the event of an earth fault?

2. In a TT system what does the first letter T represent?

3. In a TN-S system what does the third letter S represent?

4. With which type of system is the PME associated?

5. List the disconnection times for lighting and socket outlet circuits in domestic premises.

6. What reference methods should be used for cables in thermal insulation?

7. Explain the formula for calculating voltage drop.

8. Describe the earth-fault loop impedance.

9. Explain the term overload.

10. Why must you always have a fuse in a circuit?

11. Why does a circuit breaker have both thermal and magnetic devices?

12. What is the type of MCB required for protecting a domestic installation and why?

13. What value of impedance is likely to be recorded with a short circuit?

14. Explain the effects of a short circuit current.

15. What is meant by the term 'discrimination between protective devices'?

16. What type of protective device would normally have a fusing factor of around 1.8–2.0?

Chapter 3

Lighting

OVERVIEW

Illumination by means of electricity has been available for over 100 years. In that time it has changed in many ways, though many of the same ideas are still in use. The first type of elcctric lamp was the 'arc lamp', which used electrodes to draw an electrode through the air, this is now known as discharge lighting. This was quite an unsophisticated use of electricity, and many accidents and fires resulted from it. Regulations had to be developed to control discharge lighting installations, and it is interesting that the first edition of the Regulations, introduced in 1882, was entitled 'Rules and Regulations for the Prevention of Fire Risks Arising from Electric Lighting'.

This chapter will cover:

- **Incandescent lamps**
- **Regulations concerning lighting circuits**
- **Discharge lighting**

Incandescent lamps

In this method of creating light a fine filament of wire is connected across an electrical supply, which makes the filament wire heat up until it is white-hot and gives out light. The filament wire reaches a temperature of 2500–2900°C. These lamps are very inefficient and only a small proportion of the available electricity is converted into light; most of the electricity is converted into heat as infrared energy. The light output of this type of lamp is mainly found at the red end of the visual spectrum, which gives an overall warm appearance.

Operation of GLS lamps

The General Lighting Service (GLS) lamp is one type of incandescent lamp and is commonly referred to as the 'light bulb'. It has at its 'core' a very thin tungsten wire that is formed into a small coil and then coiled again.

A current is passed through the tungsten filament, which causes it to reach a temperature of 2500°C or more so that it glows brightly. At these temperatures, the oxygen in the atmosphere would combine with the filament to cause failure, so all the air is removed from the glass bulb and replaced by gases such as nitrogen and argon. Nitrogen is used to minimise the risk of arcing and argon is used to reduce the evaporation process. On low-power lamps such as 15 and 25 watt, the area inside the bulb remains a vacuum. The efficiency of a lamp is known as the efficacy. It is expressed in lumen per watt, lm/w. For this type of lamp the efficacy is between 10 and 18 lumens per watt. This is low compared with other types of lamp, and its use is limited. However, it is the most familiar type of light source used and has many advantages including:

coiled coil tungsten filament

glass bulb filled with nitrogen and argon

lead wire

exhaust tube

fuse sleeve

exhaust tube seal

cement

lead wire

soldered contacts

cap contact insulator

Figure 3.1   GLS lamp

- comparatively low initial costs
- immediate light when switched on
- no control gear
- it can easily be dimmed.

When a bulb filament finally fails it can cause a very high current to flow for a fraction of a second – often sufficient enough to operate a 5 or 6 amp miniature circuit breaker which protects the lighting circuit. High-wattage lamps, however, are provided with a tiny integral fuse within the body of the lamp to prevent damage occurring when the filament fails.

If the lamp is run at a lower voltage than that of its rating this results in the light output of the lamp being reduced at a greater rate than the electricity used by the lamp, and the lamp's efficacy is poor. This reduction in voltage, however, increases the lifespan and can be useful where lamps are difficult to replace or light output is not the main consideration.

It has been calculated that an increase in 5 per cent of the supply voltage can reduce the lamp life by half. However, if the input voltage is increased by just 1 per cent this will produce an increase of 3.5 per cent in lamp output (lumens). When you consider that the Electricity Supply Authority is allowed to vary its voltage up to and including 6 per cent it is easy to see that if this was carried on for any length of time the lamps would not last very long .

Did you know?

The average life of this type of lamp is 1,000 hours, after which the filament will rupture

Tungsten halogen lamps

These types of lamps were introduced in the 1950s. For their operation the tungsten filament is enclosed in a gas-filled quartz tube together with a carefully controlled amount of halogen such as iodine. Figure 3.2 illustrates the linear tungsten halogen lamp.

Did you know?

Running the lamp at a higher voltage than it was designed to do results in a shorter lamp life

Figure 3.2   Linear tungsten lamp

Operation of tungsten halogen lamps

The inclusion of argon and iodine in the quartz tube allows the filament to burn at a much higher temperature than the incandescent lamp. The inclusion of the halogen gas produces a regeneration effect which prolongs the life of the lamp.

As small particles of tungsten fall away from the filament, they combine with the iodine passing over the face of the quartz tube, forming a new compound. Convection currents in the tube cause this new compound to rise, passing over the filament. The intense heat of the filament causes the compound to separate into its component parts, and the tungsten is deposited back on the filament.

The lamp should not be touched with bare fingers as this would deposit grease on the quartz glass tube; this would lead to small cracks and fissures in the tube when the lamp heats rapidly, causing the lamp to fail. If accidentally touched on installation, the lamp should be cleaned with methylated spirit before being used.

44mm

16mm

6.35mm

Figure 3.3 Single-ended filament lamp

The linear type of lamp must be installed within 4° of the horizontal to prevent the halogen vapour migrating to one end of the tube, causing early failure. These types of lamps have many advantages, which include:

- increased lamp life (up to 2000 hours)
- increase in efficacy (up to 23 lumens per watt)
- reduction in lamp size.

There are two basic designs that have been produced: the double-ended linear lamp and the single-ended lamp, with both contacts embedded in the seal at one end (see Figure 3.3). This type of lamp has been produced to work on extra-low voltages: they are used extensively in the automobile industry for vehicle headlamps. They may also be used for display spotlights where extra-low voltage is desirable. They may be supplied from an in-built 230 volt/12 volt transformer.

Lamp caps

To connect the lamps to the supply there are several different caps in use. These can be divided into three groups: bayonet-fitting, screw-fitting, and plug-fitting. In general the lamp number is the diameter of the cap or the distance between pins, in millimetres. The Goliath Edison Screw (GES) design is reserved for lamps of a higher wattage such as 300 and 500 watt lamps used for floodlighting purposes.

FAQ

Q   Why does a MCB trip when a GLS lamp stops working?

A   In the cheaper lamps there are no internal fuses; therefore, what sometimes happens is that when the filament eventually burns out, some of the filament falls way, the remaining ends touch together, completing the circuit, and current flows. The filament now has a much lower resistance, which allows more current to flow, and the lamp often appears to be much brighter. The raised level of current causes the MCB to trip.

Q   Where would I use a tungsten halogen luminaire?

A   These lamps are much brighter than a GLS lamp and are therefore used where a cheap source of area floodlighting is required – frequently in conjunction with a PIR motion detector. Modern downlighters, either extra low voltage or low voltage, are now commonplace in both commercial and domestic installations.

Regulations concerning lighting circuits

Although a number of Regulations have been covered already in this book, there are several others that are applicable to the installation of lighting circuits:

Regulations concerning lighting circuits

(i) Where flexible conductors enter a luminaire as, for example, when a bulkhead fitting or batten lampholder is used, the conductors should be able to withstand any heat likely to be encountered or sleeved with heat resistant sleeving: see Regulation 522-02-02.

(ii) A ceiling rose, unless specially designed for the purpose, should have only one flexible cord: see Regulation 553-04-03.

(iii) A ceiling rose shall not be installed in any circuit operating at a voltage normally exceeding 250 volts: see Regulation 553-04-02.

(iv) Where a flexible cord supports or partly supports a luminaire, the maximum mass supported shall not exceed the values stated in Appendix 4 Table 4H3A of BS 7671.

(v) For circuits on a TN or a TT system, the outer contact of every Edison screw or single-centre bayonet cap-type lamp holder shall be connected to the neutral conductor: see Regulation 553-03-04.

(vi) Semi-conductors used in dimmer controls may be used for functional switching (not isolators) provided that they comply with Sections 512 and 537 of BS 7671.

(vii) When installing lighting circuits, the current assumed is equivalent to the connected load, with a minimum of 100 watts per lampholder: see Table 1A of the *IEE On-Site Guide*. However, it should be noted that diversity could be applied to lighting circuits in accordance with Table 1B of the *IEE On-Site Guide*.

(viii) Final circuits for discharge lighting (this includes fluorescent luminaires) shall be capable of carrying the total steady current. Where this information is not available the demand in volt-amperes can be worked out by multiplying the rated watts by 1.8: see Table 1B of the *IEE On-Site Guide*.

Figure 3.4 Lighting circuit Regulations

Discharge lighting

This is a term that refers to illumination derived from the ionisation of gas. This section looks at:

- low-pressure mercury vapour lamps
- the glow-type starter circuit
- semi-resonant starting
- high frequency
- stroboscopic effect
- other methods of starting the fluorescent tube
- other discharge lamps.

Low-pressure mercury vapour lamps

Figure 3.5 Detail of one end of a fluorescent tube

The fluorescent lamp, or, more correctly, the low-pressure mercury vapour lamp, consists of a glass tube filled with a gas such as krypton or argon and a measured amount of mercury vapour. The inside of the glass tube has a phosphor coating, and at each end there is a sealed set of oxide-coated electrodes, known as cathodes.

When a voltage is applied across the ends of a fluorescent tube the cathodes heat up, and this forms a cloud of electrons, which ionise the gas around them. The voltage to carry out this ionisation must be much higher than the voltage required to maintain the actual discharge across the lamp. Manufacturers use several methods to achieve this high voltage, usually based on a **transformer** or **choke**. This ionisation is then extended to the whole length of the tube so that the arc strikes and is then maintained in the mercury, which evaporates and takes over the discharge. The mercury arc, being at low pressure, emits little visible light but a great deal of ultraviolet, which is absorbed by the phosphor coating and transformed into visible light.

The cathodes are sealed into each end of the tube and consist of tungsten filaments coated with an electron-emitting material. Larger tubes incorporate cathode shields – iron strips bent into an oval shape to surround the cathode. The shield traps material given off by the cathodes during the tube life and thereby prevents the lamp-ends blackening.

Did you know?

The type of phosphorus used on the lamp's inner surface will determine the colour of light given out by the lamp

The gas in standard tubes is a mercury and argon mix, although some lamps (the smaller ones and the new slim energy-saving lamps) have krypton gas in them. The phosphor coating is a very important factor affecting the quantity and quality of light output. When choosing different lamps there are three main areas to be considered:

- lamp efficacy
- colour rendering
- colour appearance.

Lamp efficacy

This refers to the lumen output for a given wattage. For fluorescent lamps this varies between 40 and 90 lumens per watt.

Colour rendering

This describes a lamp's ability to show colours as they truly are. This can be important depending on the building usage. For example, it would be important in a paint shop but less so in a corridor of a building. The rendering of colour can affect people's attitude to work etc. – quite apart from the fact that in some jobs true colour may be essential. By restoring or providing a full colour range the light may also appear to be better or brighter than it really is.

Colour appearance

This is the actual look of the lamp, and the two ends of the scale are warm and cold. These extremes are related to temperatures: the higher the temperature, the cooler the lamp. This is important for the overall effect, and generally warm lamps are used to give a relaxed atmosphere while cold lamps are used where efficiency and businesslike attitudes are the priorities. The subject of lamp choice has become very complicated, and a programme of lamp rationalisation has begun. The intention is that the whole range currently available will be reduced. Also, new work has resulted in lamps with high lumen outputs and good colour rendering possibilities.

The glow-type starter circuit

In the starter, a set of normally open contacts is mounted on bi-metal strips and enclosed in an atmosphere of helium gas. When switched on, a glow discharge takes place around the open contacts in the starter, which heats up the bi-metal strips, causing them to bend and touch each other. This puts the electrodes at either end of the fluorescent tube in circuit and they warm up, giving off a cloud of electrons; simultaneously an intense magnetic field is building up in the choke, which is also in circuit.

The glow in the starter ceases once the contacts are touching so that the bi-metal strips now cool down and they spring apart again. This momentarily breaks the circuit, causing the magnetic field in the choke to rapidly collapse. The high back-e.m.f. produced provides the high voltage required for ionisation of the gas and

Did you know?

The most economic tube life is limited to around 5,000 or 6,000 hours. In industry, tubes are changed at set time intervals and all the tubes, whether still working or not, are replaced. This saves money on maintenance, stoppage of machinery and scaffolding erection etc.

Remember

The capacitor connected across the supply terminals is to correct the poor power factor that has been created by this type of light enables the main discharge across the lamp to take place. The voltage across the tube under running conditions is not sufficient to operate the starter and so the contacts remain open.

The resistance of the ionised gas gets lower and lower as it warms up and conducts more current. This could lead to disintegration of the tube. However, the choke has a secondary function as a current-limiting device: the impedance of the choke limits the current through the lamp, keeping it in balance. This is one reason why it is often referred to as ballast.

This type of starting may not succeed first time and can result in the characteristic flashing on/off, when initially switching it on.

Figure 3.6   Glow-type starter circuit

Semi-resonant starting

In this circuit the place of the **choke** is taken by a specially wound transformer. Current flows through the primary coils to one cathode of the lamp and then through the secondary coil, which is wound in opposition to the primary coil. A fairly large **capacitor** is connected between the secondary coil and the second cathode of the lamp (the other end of which is connected to the neutral).

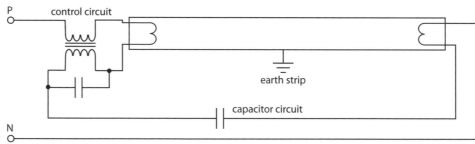

Figure 3.7   Semi-resonant circuit

The current that flows through this circuit heats the cathodes and, as the circuit is predominantly capacitive, the pre-start current leads the main voltage. Owing to the primary and secondary windings being in opposition, the voltages developed across them are 180° out of phase, so that the voltage across the tube is increased, causing the arc to strike. The primary windings then behave as a choke, thus stabilising the current in the arc. The circuit has the advantage of high power factor and easy starting at low temperatures.

High frequency

Standard fluorescent circuits operate on a mains supply frequency of 50 Hz; however, high-frequency circuits operate on about 30,000 Hz. There are a number of advantages to high frequency:

- higher lamp efficacy
- first-time starting
- noise-free
- the ballast shuts down automatically on lamp failure.

The higher efficacy for this type of circuit can lead to savings of at least 10 per cent, and in some large installations they may be as high as 30 per cent. Many high-frequency electronic ballasts will operate on a wide range of standard fluorescent lamps. The high-frequency circuit will switch on the lamp within 0.4 of a second and there should be no flicker (unlike glow-type starter circuits). Stroboscopic effect is not a problem as this light does not flicker on/off due to its high frequency.

However, a disadvantage of this type of circuit is that supply cables installed within the luminaire must not run adjacent to the leads connected to the ballast output terminals as interference may occur. Also, the initial cost of these luminaires is greater than traditional glow-type switch starts.

Stroboscopic effect

A simple example of this effect is that, when watching the wheels rotate on a horse-drawn cart on television, it appears that the wheels are stationary or even going backwards. This phenomenon is brought about by the fact that the spokes on the wheels are being rotated at about the same number of revolutions per second as the frames per second of the film being shot. This effect is known as the 'stroboscopic effect' and can also be produced by fluorescent lighting.

The discharge across the electrodes is extinguished 100 times per second, producing a flicker effect. This flicker is not normally observable but it can cause the stroboscopic effect, which can be dangerous. For example, if rotating machinery is illuminated from a single source it will appear to have slowed down, changed direction of rotation or even stopped. This is a potentially dangerous situation to any operator of rotating machines in an engineering workshop.

However, this stroboscopic effect can be harnessed – for example, to check the speed of a CD player or the speed of a motor vehicle for calibration purposes.

By using one of the following methods, the stroboscopic effect can be overcome or reduced. The first three maintain light falling on the rotating machine. The fourth makes the effective flicker at a different frequency from the operating frequency of the machines.

Did you know?

Certain frequencies of stroboscopic flash can induce degrees of drowsiness, headaches, eye fatigue and, in extreme cases, disorientation

(i) Tungsten filament lamps can be fitted locally to lathes, pillar-drilling machines etc. This will lessen the effect but will not eliminate it completely.

(ii) Adjacent fluorescent fittings can be connected to different phases of the supply. Because in a three-phase supply the phases are 120° out of phase with each other, the light falling on the machine will arrive from two different sources. Each of these will be flickering at a different time and will interfere with each other, reducing the stroboscopic effect.

(iii) Twin lamps can be wired on lead-lag circuits, thus counteracting each other. The lead-lag circuit, as the name implies, is a circuit that contains one lamp in which the **power factor** leads the other – hence the other lags. Using the leading current effect of a capacitor and the lagging current effect of an inductor produces the lead-lag effect. The lagging effect is produced naturally when an inductor is used in the circuit as shown in Figure 3.8. The leading effect uses a series capacitor, which has a greater effect than the inductor in the circuit. When these two circuits are combined as shown there is no need for further power factor correction as one circuit will correct the other.

(iv) The use of high-frequency fluorescent lighting reduces the effect by about 60 per cent.

lead circuit lamp

lag circuit lamp

series capacitor

Figure 3.8   Lead lag circuit

Other methods of starting the fluorescent tube

Quick start

The electrodes of this type of circuit are rapidly pre-heated by the end windings of an autotransformer so that a quick start is possible. The method of ionisation of the gas is the same as in the semi-resonant circuit. Difficulties may occur in starting if the voltage is low.

Thermal starter circuit

This type of circuit has waned in popularity over the years. However, there are still thousands of these fittings in service, so it is worth describing them. In this starter, the normally closed contacts are mounted on a bi-metal strip. A small heater coil heats one of these when the supply is switched on. This causes the strip to bend and the contacts to open, creating the momentary high voltage and starting the circuit discharge. The starter is easily recognised as it has four pins instead of the usual two, the extra pins being for the heater connection.

Other discharge lamps

- high-pressure mercury vapour

- low-pressure sodium

- high-pressure sodium.

Typically these consist of a glass tube known as an arc tube, in which an electrode has been fitted at either end. Rare metals such as lanthanum, lutetium or mercury may be added to the tube. The tube is pressurised with gases such as argon and neon before sealing. Lamps such as these will not normally start using a raw mains voltage but require an external igniter to provide a high-voltage injection in order to set up an electromagnetic field. This excites ions present in the gas, causing them to collide with molecules of gas, resulting in the emission of visible light. This may take some time to occur and is a characteristic of this type of lamp.

On the job: Fluorescent fault

Your employer asks you to visit the owners of an engineering manufacturing factory to discuss a problem with them. Staff have reported that some rotating machines in the main factory area seem to be slowing down. The manager has investigated this and in actual fact they are not. The main factory area is lit by single tube fluorescent lighting fittings hung from chains.

1   How would you go about investigating the problem?

2   What do you think the problem might be?'

(A suggested answer can be found on page 314)

FAQ

Q Where would I use a high-pressure mercury vapour fitting?

A High-pressure mercury vapour gives a high-quality light output, with high levels of colour rendering. Therefore it is used for lighting sports events, such as evening football and cricket matches.

Q What does it mean if, when turned on, the ends of a glow-start fluorescent tube glow but it does not strike?

A The contacts of the glow-type starter have shorted together and are not opening – if they did, the tube would strike. Replace the starter.

Q How can I tell the difference between a high-pressure sodium lamp and a high-pressure mercury vapour lamp when they **are** illuminated?

A By the colour: the high pressure sodium lamp gives of a warm golden-coloured light, while the high-pressure mercury lamp gives a blue/green light.

Knowledge check

1. List four advantages and one disadvantage of a GLS lamp.

2. What is the typical temperature at which GLS lighting will eventually settle after being switched on?

3. Which inert gases are used in GLS lamps?

4. State the efficacy and average life of a linear tungsten halogen lamp.

5. State three methods of overcoming stroboscopic effects of discharge lighting.

6. Give one reason why a fluorescent type luminaire is slow to start.

Chapter 4

Circuits and systems

OVERVIEW

This chapter will look at various circuits that are common in electrical installation work. It will look at how electricity circulates around domestic premises and how it is used in heating, security and control circuits.

This chapter will cover the following areas:

- **Timers and programmers**
- **Emergency lighting, fire alarms and standby power supplies**
- **Water heating**
- **Space heating**
- **Cooker thermostats and controllers**
- **Closed-circuit television and camera systems**
- **Intruder alarm systems**

Timers and programmers

Timers and programmers come in a wide variety of shapes and sizes and can control simple and/or complex switching operations in all aspects of modern living. Simple timers include:

- plug-in clock timers that switch lights or heaters on and off
- timers that delay the operation of contactors in motor control circuits
- timers that switch Christmas lights on and off rapidly or in a pattern.

More complex timers can be programmed to carry out multiple functions when external inputs are received or a certain time is reached. One of the most common programmers that most of you will see each day is the one used to switch the heating and hot water on at home. It is mains operated, and in the event of a power failure a rechargeable battery takes over and maintains the settings for up to two days.

These programmers can be set to switch on/off either heating or hot water or both at several different times each day, seven days per week. The component parts of a typical central heating/hot-water system consist of a multi-way connection box, circulating pump, boiler, room thermostat, hot-water tank thermostat, changeover/diverter valve and, of course, the programmer itself.

The control circuits for these are many and varied depending on the type of heating system installed. For typical control circuits the use of Danfoss Randall Control Packs with WB12 wiring centres is recommended. The WB12 includes terminal-to-terminal wiring details for Heatshare (HSP) and Heatplan (HPP) Packs.

The supply to the programmer, and hence the rest of the circuit, is usually via a fused spur with a 3 A fuse fitted. The various cables are installed to the respective component parts via the central connecting block and the programmer itself.

Figure 4.1 Wiring diagram for timer and programming domestic system (continued overleaf)

| Fully pumped system
Two 2-port zone valves with aux. sw.
PUMPED switch selection | Fully pumped system for two heating zones with domestic hot water controlled from either zone.
PUMPED switch selection on both units. |

Figure 4.1 Wiring diagram for timer and programming domestic system (continued)

When the clock part of the programmer switches on to, say, central heating, then the supply is fed via the room thermostat to the boiler and the pump until the thermostat switches itself – and thus the heating – off. The same principle applies when hot water is selected, only this time the thermostat on the hot-water cylinder cuts out.

If both heating and hot water are selected, then a diverter valve switches the flow of water via a motorised valve (one part of the system is closed off) within the system to prioritise the heating first; when that reaches temperature the diverter valve changes over and prioritises the hot-water side of the system.

On the job: Customer technical query

You have been asked to install an additional light fitting in the kitchen of a house. The central heating in the house is controlled through a central programmer located in the kitchen and an electronic digital thermostat mounted on a wall in the entrance hall. The owners ask how the electronic thermostat works with the programmer.

1   What could you tell them?

(A suggested answer can be found on page 315)

Emergency lighting, fire alarms and standby power supplies

This section looks at the following emergency lighting systems, fire alarms and standby power supplies:

- emergency lighting (maintained, non-maintained and sustained)
- fire-alarm systems
- standby power supplies.

Emergency lighting

Emergency lighting is not required in private homes because the occupants are familiar with their surroundings. However, in public buildings, people are in unfamiliar surroundings and in an emergency they will require a well-illuminated and easily identified exit route.

Emergency lighting should be planned, installed and maintained to the highest standards of reliability and integrity, so that it will operate satisfactorily when called into action. It must be installed in accordance with the British Standard Specification BS 5266: Part 1: 1999 – Code of Practice for Emergency Lighting.

Emergency-lighting terminology

For the purposes of the European Standard EN 1838, emergency lighting is regarded as a general term. There are actually several types, as shown in the diagram below:

Figure 4.2 Specific forms of emergency lighting

Emergency-escape lighting: provided to enable safe exit in the event of failure of the normal supply.

Standby-lighting: provided to enable normal activities to continue in the event of failure of the normal mains supply.

Escape-route lighting: provided to enable safe exit for occupants by providing appropriate visual conditions and direction-finding on escape routes and in special areas/locations, and to ensure that fire-fighting and safety equipment can be readily located and used.

Open area (or anti-panic area) lighting: provided to reduce the likelihood of panic and to enable safe movement of occupants towards escape routes by providing appropriate visual conditions and direction-finding.

High-risk-task area lighting: provided to ensure the safety of people involved in a potentially dangerous process or situation and to enable proper shut-down procedures to be carried out for the safety of other occupants of the premises.

Types

Emergency lighting comes in two main formats: individual, self-contained systems with their own emergency battery power source, and centralised battery-backup systems. In using both these formats there are then three types available:

Maintained: The same lamp is used by both the mains and the emergency backup system and therefore operates continuously. The lamp is supplied by an alternative supply when the mains supply fails.

The advantage of this system is that the lamp is continuously lit and therefore we can see whether a lamp needs replacing. The disadvantage is that, although the lamp is lit, we do not know whether it is being powered by the mains supply or the batteries. It is therefore common to find a buzzer and indicator lamp that show which supply is being used. Emergency lighting should be of the maintained type in areas in which the normal lighting can be dimmed, e.g. theatres or cinemas, or where alcohol is served.

Non-maintained: The emergency lighting lamp only operates when the normal mains lighting fails. Failure of the mains supply connects the emergency lamps to the battery supply. The disadvantage of this system is that a broken lamp will not be detected until it is required to operate. It is therefore common to find an emergency-lighting test switch available that disconnects the mains supply for test purposes.

Sustained: An additional lamp housed in the mains luminaire is used only when the mains fails.

The duration of the emergency lighting is normally three hours in places of entertainment and for sleeping risk, or where evacuation is not immediate, but one hour's duration may be acceptable in some premises if evacuation is immediate and re-occupation is delayed until the system has recharged.

Siting of luminaires

BS 5266 and IS 3217 provide detailed guidance on where luminaires should be installed and what minimum levels of illuminance should be achieved on escape routes and in open areas. It also specifies what minimum period of duration should be achieved after failure of the normal mains lighting.

Local and national statutory authorities, using legislative powers, usually require escape lighting. Escape-lighting schemes should be planned so that identifiable features and obstructions are visible in the lower levels of illumination that will occur during an emergency.

Current UK regulations require the provision of a horizontal illuminance at floor level on the centre line of a defined escape route, of not less than 0.2 lux (similar to the brightness of a full moon). In addition, for escape routes of up to 2 m wide, 50 per cent of the route width should be lit to a minimum of 0.1 lux. Wider escape routes can be treated as a number of 2 m wide bands.

Emergency-escape lighting should:

- indicate the escape routes clearly, allowing for changes of direction or of level
- provide illumination along escape routes to allow safe movement towards the final exits
- ensure that fire-alarm call points and fire-fighting equipment can be readily located.

Standby lighting is required in, for example, hospital operating theatres and in industry, where an operation or process, once started, must continue even if the mains lighting fails.

Additional emergency lighting should also be provided in:

- lift cars – potential for the public to be trapped
- toilet facilities – particularly disabled toilets – and open tiled areas over 8 m²
- escalators – to enable users to get off them safely
- motor generator, control or plant rooms. These require battery-supplied emergency lighting to help any maintenance or operating personnel
- covered car parks along the normal pedestrian routes.

Illuminance levels for open areas

Emergency lighting is required for areas larger than 60 m² or open areas with an escape route passing through. Illuminance BS 5266 requires 1 lux average over the floor area. The European standard EN 1838 requires 0.5 lux minimum anywhere on the floor level excluding the shadowing effects of contents. The core area excludes the 0.5 m to the perimeter of the area.

High-risk-task area lighting

BS 5266 requires that higher levels of emergency lighting are provided in areas of particular risk, although no values are defined. The European standard EN 1838 says that the average horizontal illuminance on the reference plane (note that this is not necessarily the floor) should be as high as the task demands in areas of high risk. It should not be less than 10 per cent of the normal illuminance, or 15 lux, whichever is the greater. It should be provided within 0.5 seconds and continue for as long as the hazard exists. This can normally only be achieved by a tungsten or permanently

Did you know?

Cashpoints in commercial buildings need to be illuminated at all times to discourage acts of theft occurring during a mains failure illuminated maintained fluorescent lamp source. The required illuminance can often be achieved by careful location of emergency luminaires at the hazard, and may not require additional fittings.

Maintenance

Essential servicing should be defined to ensure that the system remains at full operational status. This would normally be performed as part of the testing routine, but for consumable items, such as replacement lamps, spares should be provided for immediate use.

Fire-alarm systems

A correctly installed fire-alarm system installation is of paramount importance and can be compared to any other electrical undertaking, as life could be lost and property damaged as a result of carelessly or incorrectly connected fire-detection and alarm equipment. The subject is detailed, and therefore this section sets out only to give an overview of requirements.

BS 5839 Part 1 classifies fire-alarm systems, perhaps better described as fire-detection and alarm systems, into the following general types:

Did you know?

The Building Regulations require the installation of mains-fed smoke detectors in new-build domestic installations

- Type M: Break-glass contacts operating sounders for protection of life. No automatic detection
- Type L: Automatic detection systems for the protection of life
- Type P: Automatic detection systems for the protection of property.

It is essential that the installation of fire-alarm systems is carried out in compliance with the requirements of BS 5839 Part 1, BS 7671 and manufacturers' instructions, but remember: local government can enforce even stricter requirements in the interests of public safety. BS 5839 and BS 7671 (528-01-04) state that fire-alarm circuits must be segregated from other circuits, and in order to comply with BS 7671 a dedicated circuit must be installed to supply mains power to the fire-alarm control panel.

Fire-alarm systems can be designed and installed for one of two reasons:

- property protection
- life protection.

Property protection

A satisfactory fire-alarm system for the protection of property will automatically detect a fire at an early stage, indicate its location and raise an effective alarm in time to summon fire-fighting forces (both the resident staff and the fire service). The general attendance time of the fire service should be less than 10 minutes. Therefore an automatic direct link to the fire service is a normal part of such a system.

A fire-alarm system might have prevented this

Protection for property is classed as:

- P1: All areas of the building must be covered with detectors with the exception of lavatories, water closets and voids less than 800 mm in height, such that spread of fire cannot take place in them prior to detection by detectors outside the void.

- P2: Only defined areas of high risk are covered by detectors. A fire-resisting construction should separate unprotected areas.

Life protection

A satisfactory fire-alarm system for the protection of life can be relied upon to sound a fire alarm in sufficient time to enable the occupants to escape. Life protection is classed as:

- M: The most basic and minimum requirement for life protection. It relies upon manual operation of call points and therefore requires people to activate the system. Such a system can be enhanced to provide greater cover by integrating any, or a combination, of the following:

- L1: Same as P1 (as previously mentioned)

- L2: Only provides detection in specified areas where a fire could lead to a high risk to life, e.g. sleeping areas, kitchens, day accommodation etc., and places where the occupants are especially vulnerable owing to age or illness or are unfamiliar with the building. An L2 system always includes L3 coverage.

- L3: Protection of escape routes. The following areas should therefore be included:
  - (i) corridors, passageways and circulation areas
  - (ii) all rooms opening onto escape routes
  - (iii) stairwells
  - (iv) landing ceilings
  - (v) the top of vertical risers, e.g. lift shafts
  - (vi) at each level within 1.5 m of access to lift shafts or other vertical risers.

Types of fire-alarm system

All fire-alarm systems operate on the same general principle, i.e. if a detector detects smoke or heat or if a person operates a break-glass contact, then the alarm will sound. We will look at the devices that may be incorporated into the system later. That said, most fire-alarm systems belong to one of the following categories:

Conventional

In this type of system, a number of devices (break-glass contacts/detectors) are wired as a radial circuit from the control panel to form a zone (e.g. one floor of a building). The control panel would have lamps on the front to indicate each zone, and if a device operates then the relevant zone lamp would light up on the control panel. However, the actual device that has operated is not indicated.

Identifying accurately where the fire has started would therefore depend on having a number of zones and knowing where in the building each zone is. Such systems are therefore normally found in smaller buildings or where a cheap, simple system is required.

Addressable

The basic principle here is the same as for a conventional system, the difference being that by using modern technology the control panel can identify exactly which device initiated the alarm.

These systems have their detection circuits wired as loops, with each device then having an 'address' built in. Such systems therefore help fire location by identifying the precise location of an initiation, and thus allow the fire services to get to the source of a fire more quickly.

Radio addressable

These are the same as addressable systems, but have the advantage of being wireless and can thus reduce installation time.

Analogue

Sometimes known as intelligent systems, analogue systems incorporate more features than either conventional or addressable systems. The detectors may include their own mini-computer, and this evaluates the environment around the detector and is therefore able to let the control panel know whether there is a fire, a change in circumstance likely to lead to a fire, a fault, or even if the detector head needs cleaning. Consequently these systems are useful in preventing the occurrence of false alarms.

Fire-prevention systems

Although still incorporating fire-detection systems, one recent innovation has been the introduction of the fire-reduction system. This type of system is still under development, but works by reducing levels of one of the main components in the fire triangle – oxygen – and thus seeks to create a 'fire-free' area. Although not without problems, usage of these systems could be appropriate in critical areas such as historical archives or identified unmanned areas such as chemical storage.

Zones

To ensure a fast and unambiguous identification of the source of fire, the protected area should be divided into zones. Although less essential in analogue addressable systems, the following guidelines relate to zones:

- If the floor area of each building is not greater than $300\,m^2$ then the building only needs one zone, no matter how many floors it has. This covers most domestic installations.

- The total floor area for one zone should not exceed $2,000\,m^2$.

- The search distance should not exceed 30 m. This means the distance that has to be travelled by a searcher inside a zone to determine visually the position of a fire should not be more than 30 m. The use of remote indicator lamps outside of doors may reduce the number of zones required.

- Where stairwells or similar structures extend beyond one floor but are in one fire compartment, the stairwell should be a separate zone.

- If the zone covers more than one fire compartment then the zone boundaries should follow compartment boundaries.

- If the building is split into several occupancies, no zone should be split between two occupancies.

System devices

The control panel

This is the heart of any system, as it monitors the detection devices and their wiring for faults and operation. If a device operates, the panel operates the sounders as well as any other related equipment and gives an indication of the area in which the alarm originated.

Break-glass contacts (manual call points)

The break-glass call point is a device to enable personnel to raise the alarm in the event of a fire, by simply breaking a fragile glass cover (housed in a thin plastic membrane to protect the operative from injury sustained by broken or splintered glass). A sturdy thumb pressure is all that is required to rupture the glass and activate the alarm. The following guidance relates to the correct siting and positioning of break-glass call points:

- They should be located on exit routes and in particular on the floor landings of staircases and at all exits to the open air.

- They should be located so that no person need travel more than 30 m from any position within the premises in order to raise the alarm.

- Generally, call points should be fixed at a height of 1.4 m above the floor, at easily accessible, well-illuminated and conspicuous positions free from obstruction.

- The method of operation of all manual call points in an installation should be identical unless there is a special reason for differentiation.

- Manual and automatic devices may be installed on the same system, although it may be advisable to install the manual call points on separate zones for speed of identification.

Automatic detectors

When choosing the type of detector to be used in a particular area it is important to remember that the detector has to discriminate between fire and the normal

Break-glass call point

environment existing within the building, for example smoking in hotel bedrooms, fumes from fork-lift trucks in warehouses, or steam from kitchens and bathrooms. There are several automatic detectors available, as described below.

Heat detectors (fixed-temperature type)

The fixed-temperature heat detector is a simple device designed to activate the alarm circuit once a predetermined temperature is reached. Usually a choice of two operational temperatures is available: either 60°C or 90°C. This type of detector is suitable for monitoring boiler-rooms or kitchens where fluctuations in ambient temperature are commonplace.

Heat detector (rate-of-rise type)

This type of detector responds to rapid rises of temperature by sampling the temperature difference between two heat-sensitive thermocouples or thermistors mounted in a single housing (a thermistor is a device whose resistance quickly decreases with an increase in temperature).

Smoke detectors

May be either of the 'ionisation' or 'optical' type. Smoke detectors are not normally installed in kitchens, as burning toast and so on could activate the alarm.

The ionisation detector is very sensitive to smoke with fine particles such as that from burning paper or spirit, whereas the optical detector is sensitive to 'optically dense' smoke with large particles such as that from burning plastics.

The optical smoke detector, sometimes known as the photoelectric smoke detector, operates by means of the light-scattering principle. A pulsed infrared light is targeted at a photo-receiver but separated by an angled non-reflective baffle positioned across the inner chamber. When smoke and combustion particles enter the chamber, light is scattered and reflected on to the sensitive photo receiver, triggering the alarm.

Detector heads for fire-alarm systems should only be fitted after all trades have completed work, as their work could create dust, which impairs the detector operation. Strict rules exist regarding the location of smoke detectors.

Alarm sounders

These are normally either a bell or an electronic sounder, which must be audible throughout the building in order to alert (and/or evacuate) the occupants of the building. The following gives guidance for the correct use of alarm sounders:

- A minimum level of either 65 dBA, or 5 dBA above any background noise likely to persist for a period longer that 30 seconds, should be produced by the sounders at any occupiable point in the building.

- If the alarm system is to be used in premises such as hotels, boarding houses etc. where it is required to wake sleeping persons, then the sound level should be a minimum of 75 dBA at the bedhead.

Remember

It is wise not to install a rate-of-rise heat detector unit in a boiler-room or kitchen, where fluctuations in ambient temperature occur regularly. This will help to avoid nuisance alarms

Smoke detector

Alarm bell

- All audible warning devices used in the same system should have a similar sound.

- A large number of quieter sounders rather than a few very loud sounders may be preferable. At least one sounder will be required per fire compartment.

- The level of sound should not be so high as to cause permanent damage to hearing.

Wiring systems for fire alarms

BS 5839 Part 1 recommends 11 types of cable that may be used where prolonged operation of the system in a fire is not required. However, only two types of cable may be used where prolonged operation in a fire is required.

It is obvious that the cabling for sounders and any other device intended to operate once a fire has been detected must be fireproof. However, detection wiring can be treated differently as it can be argued that such wiring is only necessary to detect the fire and sound the alarm.

In reality fire-resistant cabling tends to be used throughout a fire-alarm installation for both detection and alarm wiring. Consequently, as an example, MICC cable used throughout the system is considered by many as the most appropriate form of wiring, but there are alternatives, such as Fire-tuf.

Irrespective of the cable type and the circuit arrangements of the system, all wiring must be installed in accordance with BS 7671. Where possible, cables should be routed through areas of low fire risk, and where there is risk of mechanical damage they should be protected accordingly.

Because of the importance of the fire-alarm system, it is wise to leave the wiring of the system until most of the constructional work has been completed. This will help prevent accidental damage occurring to the cables. Similarly, keep the control panel and activation devices in their packing cartons, and only remove them when building work has been completed in the area where they are to be mounted, thus preventing possible damage to the units.

Standby back-up for fire-alarm systems

The standby supply, which is usually a battery, must be capable of powering the system in full normal operation for at least 24 hours, and, at the end of that time period, must still have sufficient capacity to sound the alarm sounders in all zones for a further 30 minutes.

Find out

What other fire-resistant cabling is available?

Typical maintenance checks for a fire-alarm system

BS 5839 Part 1 makes the following recommendations:

Daily inspection	Annual test
• Check that the control panel indicates normal operation. Report any fault indicators or sounders not operating to the designated responsible person.	• Repeat the quarterly test • Check all call points and detectors for correct operation • Enter details of test in logbook.
Weekly test	**Every two to three years**
• Check panel key operation and reset button • Test fire alarm from a call point (different one each week) and check sounders • Reset fire-alarm panel • Check all call points and detectors for obstruction • Enter details of test in logbook.	• Clean smoke detectors using specialist equipment • Enter details of maintenance in logbook.
Quarterly test	**Every five years**
• Check all logbook entries and make sure any remedial actions have been carried out. • Examine battery and battery connections • Operate a call point and detector in each zone • Check that all sounders are operating • Check that all functions of the control panel are operating by simulating a fault • Check sounders operate on battery only • Enter details of test in logbook.	• Replace battery (see manufacturer's information).

Table 4.1   BS 5839 recommendations

Standby power supplies

The electrical supply in this country is very reliable and secure. However, as with all systems there are occasional interruptions which for some installations would be dangerous as well as inconvenient. Hospitals, air-traffic control and the petro-chemical industry are just a few installations that could not tolerate an interruption to the mains supply, so a standby system needs to be available.

Smaller establishments such as small offices cannot afford complex standby generation systems, but nevertheless they may have computer systems that cannot afford to be off or, worse still, risk losing data. In this situation standby power systems known as Uninterruptible Power Supplies (UPS) are used, which consist of a battery supply that is charged up via the mains when not in use. When the mains supply is lost the UPS automatically cuts in and, via the electronics contained in it, converts the d.c. battery supply to a mains supply capable of powering several computers.

Remember

Large installations need a standby generating system, whereby a large combustion engine cuts in automatically and drives a generator capable of supplying the load needed to continue working safely

Water heating

80 mm lagging jacket

outlet

thermostats

short element (day-time top-up)

long element (night store)

inlet

Figure 4.3 Dual-element immersion heater, hot water

There are two main methods of heating water electrically: either heating a large quantity stored in a tank or heating only what is required when it is needed. With both of these types of heater it is important to ensure that the exposed and extraneous conductive parts (refer to Chapter 2 Earthing and protection) are adequately bonded to earth: water and electricity do not mix together well! It is also important to ensure that the cables selected are of the correct size for full load current, since no diversity is allowed for water heaters (refer to the Cable Selection section in Chapter 2).

Heating large tanks of stored water (typically 137+ litres) is done using an immersion heater (see Figure 4.3) fitted into a large water tank and then controlling when it is on or off via either a timer switch or an on/off switch.

The temperature of the water is controlled by a stem-type thermostat (similar to the oven type described later in the Cooker section), which is incorporated within the housing of the heating element. This type of heater is used in domestic situations, although larger multiple immersion heaters can be used in industrial situations. The heater in a domestic situation must be fed from its own fuse/MCB in the consumer unit and have a double pole isolator fitted next to the storage tank. The final connection to the heating element must be with heatproof flexible cable due to the high ambient temperatures where the water tank is normally located.

This type of system sometimes has two elements, one of which is controlled via a separate supply which only operates at night time (Economy 7 or white-meter supply) when cheap electricity is available, thus heating a full tank of water ready for use the next day; the other is switched on as and when needed during the day to boost the amount of hot water available.

The rest of this section will describe the different types of water system:

- cistern type
- non-pressure
- instantaneous.

Cistern-type

Where larger volumes of hot water are needed, for example in a large guest house, then a cistern-type water heater (9 kW+) is used which is capable of supplying enough hot water to several outlets at the same time.

Figure 4.4 Cistern-type water heater

Non-pressure

Non-pressure water heaters, which are typically rated at less than 3 kW and contain less than 15 litres of water, heat the water ready for use and are usually situated directly over the sink, such as in a small shop or hairdresser's salon.

Figure 4.5 Non-pressure water heater

Instantaneous

Instantaneous water heaters heat only the water that is needed. This is done by controlling the flow of water through a small internal water tank which has heating elements inside it; the more restricted the flow of water then the hotter the water becomes.

The temperature of the water can therefore be continuously altered or stabilised locally at whatever temperature is selected. This is how an electric shower works, and showers in excess of 10 kW are currently available. The shower-type water heater must be supplied via its own fuse/MCB in the consumer unit and have a double pole isolator located near the shower.

Figure 4.6 Instantaneous water heater

Space heating

The type of electric heating available falls into two main categories: direct acting heaters and thermal storage devices.

Direct acting heaters

Direct acting heaters are usually just switched on and off when needed; some of them can be thermostatically controlled. Direct heaters fall into two categories: radiant and convection.

Radiant heaters

The radiant-type heaters reflect heat and come in a variety of shapes, sizes and construction as follows.

Traditional electric fire: has a heating element supported on insulated blocks with a highly polished reflective surface behind it; these range in size from about 750W to 3kW.

Infrared heater: consists of an iconel-sheathed element or a nickel-chrome spiral element housed in a glass silica tube which is mounted in front of a highly polished surface. Sizes vary from about 500 W to 3 kW; the smaller versions are usually suitable for use in bathrooms and may be incorporated with a bulb to form a combined heating and lighting unit.

Oil-filled radiator: consists of a pressed steel casing in which are housed heating elements; the whole unit is filled with oil. Oil is used because it has a lower specific gravity than water and so heats up and cools down more quickly. Surface temperature reaches about 70°C, and power sizes range from about 500 W to 2 kW.

Tubular heater: low-temperature unit designed to supplement the main heating in the building. Consists of a mild steel or aluminium tube of about 50 mm diameter in which is mounted a heater element. The elements themselves are rated at 200 W to 260 W per metre length and can range in length from about 300 mm up to 4.5 m. The surface temperature is approximately 88°C.

Under-floor heater: consists of heating elements embedded under the floor which heat up the tiles attached to the floor surface. The floor then becomes a large low-temperature radiant heater, and a room thermostat controls the temperature within the room. The floor temperature does not normally exceed 24°C. The elements have conductors made from a variety of materials such as chromium, copper, aluminium, silicon or manganese alloys. The insulating materials used are also made from a variety of materials such as asbestos, PVC, silicon rubber and nylon.

Convection heaters

Convection heaters consist of a heating element housed inside a metal cabinet that is insulated both thermally and electrically from the case so that the heat produced warms the surrounding air inside the cabinet. Cool air enters the bottom of the cabinet and warm air is passed out at the top of the unit at a temperature of between 80 and 90°C. A thermostatic control is usually fitted to this type of heater.

Fan heater: operates in the same way as a convector heater but uses a fan for expelling the warm air into the room. Fan heaters usually have a two-speed fan incorporated into the casing and up to 3 kW of heating elements.

Thermal storage devices

Thermal storage heaters heat up thermal blocks within the unit during off-peak times to enable use of cheap-rate electricity. The heat stored is then released during the day when it is needed.

A thermal storage unit consists of several heating elements mounted inside firebricks, which in turn are surrounded by thermal insulation such as fibreglass, all housed inside a metal cabinet. The firebricks are made from clay, olivine, chrome and magnesite, which have very good heat-retaining properties. The bricks are heated up during off-peak hours (usually less than half the normal price per electrical unit) and the heat is stored within the bricks until the outlet vent is opened the following day and allows the warm air to escape and hence heat up the room.

Cooker thermostats and controllers

This section covers:

- simmerstat

- oven thermostats.

Simmerstat (energy regulator)

This device is used to control the temperature of electric cooking plates. It uses a bi-metal strip as its main principle of operation; it is not controlled by the temperature of the hotplates. Operation is by the opening and closing of a switch at short definite time intervals by the heating up (via an internal heating coil) of a bi-metal strip. The length of time that the switch is opened or closed is determined by the control knob mounted on the front of the device, and hence the length of time that the hotplate has power is varied. The control knob is normally calibrated from either 0 to 10 or 0 to 5, with the highest number being the hottest temperature that the hotplate will reach.

There are two basic ways that these devices are arranged. One is with a shunt (parallel)-connected heating coil, and the other is with a series-connected heater coil. The heater coil in each case responds to the current flowing through it and hence determines the control of the bi-metal strip.

Figure 4.7 Shunt-wired regulator

The shunt-connected thermal regulator (Figure 4.7) consists of a two-part bi-metal strip block where one strip 'A' has a small-gauge heater wire wrapped around it; this is in turn connected in parallel with the hotplate element. The second part of the bi-metal strip block 'B' is in mechanical contact with the cam of the control knob. Both of these strips are connected together at one end and pivoted on a fulcrum point 'C'. When the control knob is in the 'off' position, the cam is pushing against the bi-metal strip and hence keeps the contacts at 'F' open, so that no current flows and therefore the hotplate does not heat up.

When the control knob is moved to one of the 'on' positions, the cam moves and pressure on the bi-metal strip is reduced, thus allowing the contacts at 'F' to close and start to heat up the hotplate. As this happens current flows through the heater coil as well as the hotplate and causes the bi-metal strip to bend, causing the contacts at 'F' to open. When the heater coil and bi-metal strip cool down the bi-metal strip bends and allows the contacts at 'F' to close again, thus repeating the cycle.

The hotplate therefore has power switched 'on' and 'off' rapidly and hence stays at a constant temperature. This technique is known as 'simmering', hence the name of the device 'simmerstat'. At low-temperature settings the contacts will be open for longer, and in the fully 'on' position the contacts at 'F' will be closed and no regulation occurs.

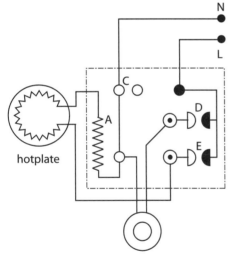

Figure 4.8 Series-wired regulator

In a series type of simmerstat as shown in Figure 4.8, 'A' is the bi-metallic heater coil, which is connected in series with the hotplate. On the right of the diagram, there are two sets of contacts, 'D' and 'E'. When the control knob is operated as if to turn the hotplate 'on', both contacts will close. One of them 'D' will bring a pilot lamp into the circuit to show that the hotplate is being heated, the other 'E' is the main contact that will energise the hotplate.

As the heater coil 'A' transmits heat it causes the bi-metal strip to bend, and eventually this will cause the contacts at 'C' to open, thus breaking the neutral to the coil and the pilot lamp. Consequently the pilot lamp will go out and the hotplate is switched 'off'.

When this happens the heater coil begins to cool and the bi-metal strip returns to its original position, allowing the contacts at 'C' to close again, and thus the cycle is repeated. Each time the contacts open, the pilot light goes off and gives a visual indication that the hotplate is up to temperature.

Oven thermostats

There are two basic types of oven thermostat, both of which work in similar ways. A capillary type has a capillary tube (typically about 800 mm long) filled with liquid which, when the phial containing it heats up, expands. The liquid then pushes against the capsule in the control housing, which in turn pushes against the plunger in the pressure block and causes the contacts to open and hence switch off the oven-heating elements.

The other type of oven housing is the stem type. These come in various lengths to suit different types of oven. The operation of the stem type relies on two dissimilar metals

Figure 4.9   Capillary type oven thermostat

Figure 4.10   Stem type oven thermostat

expanding at different rates. It consists of a copper tube with an **Invar** rod inside. When the temperature increases, the copper tube expands faster than the invar rod and pushes against the pressure rod; this opens the contacts and the oven element is switched off. When it cools down, the copper tube 'shrinks' and allows the invar plunger rod to move down and close the contact for the oven element. This type of thermostat is also used in water-heating systems.

Closed-circuit television and camera systems (CCTV)

There are many different types of CCTV systems in use today, ranging from those suitable for domestic properties through to sophisticated multi-camera/multi-screen monitoring for large commercial and industrial premises.

Typical systems

In this section we will look at the component parts that make up a typical system. The following topics will be looked at:

- wireless CCTV
- cameras
- monitoring and recording
- wired CCTV
- light levels
- other systems.

Wireless CCTV

These systems do not require cabling back to a monitor or video recorder; they have an in-built transmitter which transmits the image seen back to these pieces of equipment. Typically they can transmit 100 m outdoors and 30 m indoors. They do, however, still require a power supply (usually 9–12 V d.c.), which is usually obtained via a small power supply connected to the mains. These systems are useful where it is difficult to install video cable back to the monitor or video recorder but they can suffer from interference problems.

Wired CCTV

These systems do require cabling back to the monitor or video recorder but can be positioned many hundreds of metres away from them. Usually the same cable will provide power and the video signal back to the recording device, so all the power supplies for many cameras can be located at one central control point.

Cameras

There are many different types of camera available, ranging from very cheap (less than £100) to those costing many thousands of pounds. There are two common types: CMOS and CCD. The CMOS type is the cheapest but the images produced are not very clear or sharp. The CCD camera, on the other hand, produces very clear and sharp images from which people are easily identifiable.

Most cameras are installed outside and therefore virtually all cameras available are weatherproof; if they are not they will need to be fitted into a weatherproof housing. Virtually all cameras have the lens integrated into the camera and are sealed to prevent moisture getting in; thus they do not need a heater built in to keep the lens dry. With the lens being sealed into the camera, the former cannot be adjusted, so only one field of view is possible.

Colour and monochrome types of camera are available, with colour cameras being the more expensive. Colour cameras can only transmit colour if the light level is high, so generally speaking they will not transmit colour images at night. Monochrome cameras, on the other hand, can incorporate infrared (IR) sensitivity, allowing for clearer images where discreet IR illumination is available.

Light levels

Light levels available where the camera is to be used are an important consideration. Table 4.2 shows some typical light levels.

When choosing a suitable camera for a particular environment, it is best to select one that is specified at approximately 10 times the minimum light level for the environment. One that is specified at the same level of light will not produce the clear images needed, because the camera will not have enough light to 'see'.

Environment	Typical light level
Summer sunlight	50,000 lux
Dull daylight	10,000 lux
Shop/office	500 lux
Main street lighting	30 lux
Dawn/dusk	1–10 lux
Side street lighting	3 lux

Table 4.2 Typical light levels

Monitoring and recording

Most CCTV systems use several cameras, each relaying images back to a central control where they are either viewed or recorded. Three methods for recording or viewing these images are by using a video switcher, a quad processor or a multiplexer.

Switcher

A CCTV switcher, as the name suggests, is a device that switches between camera images one at a time. The image can either be viewed or recorded onto a video recorder; only one image at a time can be accessed.

Quad processor

This device enables four camera images to be viewed on one screen at the same time, or one image or all four to be recorded at the same time. The quality of the image when recording all four is not as good.

Multiplexer

This device allows simultaneous recording of multiple full-sized images onto one VCR, or can allow more than one camera image to be displayed at the same time without losing picture quality.

For recording purposes, a slower-moving tape can record the images for long periods of time. The time lapse can be set for either 24, 240 or 960 hours of recording on standard tapes.

Other systems

PC-based systems

By adding a video capture card and surveillance software to a PC, a powerful digital system can be created. Some of the advantages are:

- it is easy to expand the system
- it is easy to record (via hard drive)
- images can be emailed
- text alerting is possible
- software allows many configurations for monitoring
- remote viewing is possible.

Motion detectors

The camera and recording facilities are only activated when movement is detected within the camera's range. Typically this is activated by the use of passive infrared sensors (PIRs) similar to those used on security lighting and alarm systems.

Intruder alarm systems

Alarm systems are increasingly seen as standard equipment in a house or office. They act as a deterrent to some intruders but will never stop the more determined ones. People feel more secure when they have an alarm installed, and in most cases it will reduce their insurance premiums. There are basically two ways to protect a property: one is called perimeter protection and the other is space detection.

Perimeter protection detects a potential intruder before they gain entry to the premises, whereas space detection only detects when the intruder is already on the premises. Sometimes both types are used together for extra security.

Typical systems

In this section we look at some of the component parts of an alarm system and some of the more common types of detection devices available and what they do:

- proximity switches
- inertia switches
- passive infrared
- ultrasonic devices
- control panels
- audible and visual warning devices.

Proximity switch

This is a two-part device: one part is a magnet and the other contains a reed switch. The two parts are fixed side by side (usually less than 6 mm apart) on a door or window, and when the door or window is opened the reed switch opens (because the magnet no longer holds it closed) and activates the alarm panel. The switch can be surface-mounted or can be recessed into the door or window frame. This device is generally used for perimeter protection and does not rely on a power supply to operate.

Inertia switch

This type of switch detects the vibration created when a door or window is forced open. This then sends a signal to the alarm panel and activates the sounder. The sensitivity of these devices can be adjusted, and they are used for perimeter protection. These need a 12 volt d.c. supply to operate.

Passive infrared

These devices are used to protect large areas of space and are only activated when the intruder has already gained entry. The device monitors infrared which detects the movement of body heat across its viewing range; this in turn sends a signal to the panel and activates the sounder. These can be adjusted for range and, by fitting different lenses, the angle of detection can also be adjusted. These need a 12 volt d.c. supply to operate.

Ultrasonic devices

These devices send out sound waves and receive back the same waves when no-one is in the building. However, when an intruder enters the detection range, the sound waves change (because of deflection) and trigger the alarm panel. These devices also require a 12 volt d.c. supply for operation and are used for space-detection systems.

Control panels

Control panels are the 'brains' of the system to which all the parts of the system are connected. They used to be key-operated but nowadays they virtually all use a digital keypad, either on the panel itself or mounted remotely elsewhere in the building, for switching the system on or off.

The panels are all programmable whereby entry- and exit-route zone delays can be adjusted, new codes selected for switching on/off, automatic telephone diallers set to ring any phone selected etc.

Control panels have a mains supply installed which is reduced (via a transformer) down to 12 volts d.c. for operation of all the component parts that need it. A rechargeable battery backup is provided in case of mains failure.

Audible and visual warning devices

When an alarm condition occurs a means of attracting attention is obviously needed, either audibly or visually or sometimes both. The most common audible sounder is the electronic horn (I'm sure you've all heard them before!), which will sound for 20 minutes (the maximum allowed by law) before being switched off by the panel automatically. The panel then re-arms itself and monitors the system again.

To help identify which alarm has sounded (especially when there are several in the same area) a visual warning is usually fitted to the sounder box, which activates at the same time. This is a xenon light (strobe light) and can be obtained in a variety of colours. This light usually remains on after the alarm has automatically been reset to warn the occupant upon their return that an alarm condition has occurred. It is only reset when the control panel itself is reset by the occupant.

FAQ

Q Can I use PVC/PVC cable to wire in sounders?

A No, these need to be wired in a flameproof cable such as MICC, FP200 or Fire-tuf.

Q Why is it necessary to put emergency lighting outside buildings?

A To avoid panic and accidents if there is a need to evacuate a building during the hours of darkness. In the event of a power failure, people would be moving out of an illuminated area into the darkness, possibly tripping and blocking the exit for others. Putting an emergency light outside exits prevents this.

Knowledge check

1. What are the two reasons why emergency lighting is provided?

2. Describe the main difference between maintained and non-maintained emergency lighting.

3. List the segregation requirements of fire alarms to BS 5839.

4. Name three situations requiring emergency lighting.

5. List four types of automatic fire detectors.

6. State the primary reasons for zoning a fire-alarm system within a building.

7. Why is it necessary to use a fire-resistant cable on a fire-alarm sounder circuit?

8. Typical maintenance checks are carried out on fire-alarm systems. What is checked annually?

9. State three types of water heater.

10. State two types of device used in a cooker to control temperature and regulate energy.

11. Why should a CCTV camera have a minimum lux level of ten times the lux level of the location it is monitoring?

12. List the typical six component parts of an intruder alarm-system.

Chapter 5

Inspection, testing and commissioning

OVERVIEW

Electrical installations should be designed into separate circuits, e.g. power, lighting, cooker circuits, to avoid danger in the event of faults and to facilitate the safe operation of the inspection and testing process. It is a requirement of the *Electricity at Work Regulations 1989* that this information will be available as an on-site record of the design. This chapter examines the requirements for inspecting and testing an installation. It will cover the following:

- **General requirements**
- **Initial verification, certification and commissioning**
- **The sequence of initial tests**
- **Certificates**
- **Initial inspection**
- **Periodic inspection and testing**

General requirements

When to inspect and test

Initial inspection and testing is necessary on all newly completed installations. In addition, because all electrical installations deteriorate due to a number of factors such as damage, wear and tear, corrosion, excessive electrical loading, ageing and environmental influences, periodic inspection and testing must be carried out at regular intervals determined by the following:

- Legislation requires that all installations must be maintained in a safe condition and therefore must be periodically inspected and tested. Table 5.1 details the maximum period between inspections of various types of installation.

- Licensing authorities, public bodies, insurance companies and other authorities may require public inspection and testing of electrical installations.

- The installation must be checked to ensure that it complies with BS 7671.

- It is also recommended that inspection and testing of installations should occur when there is a change of use of the premises, any alterations or additions to the original installation, any significant change in the electrical loading of the installation, and where there is reason to believe that damage may have been caused to the installation.

Domestic Or if change of occupancy	10 years	Hotels	5 years
		Theatres	3 years
Commercial Or if change of occupancy	5 years	Public houses	5 years
		Village halls/community centres	5 years
Educational premises	5 years	Agricultural/horticultural	3 years
Industrial	3 years	Caravans	3 years
Hospitals	5 years	Caravan parks	1 year
Residential accommodation Or if change of occupancy	1 year	Highway power supplies	6 years
		Marinas	1 year
Offices	5 years	Fish farms	1 year
Shops	5 years	Swimming pools	1 year
Laboratories	5 years	Emergency lighting	3 years
Cinemas	3 years	Fire alarms	1 year
Churches	5 years	Launderettes	1 year
Leisure complexes	3 years	Petrol filling stations	1 year
Places of public entertainment	3 years	Construction sites	3 months
Restaurants	5 years		

Table 5.1 Frequency of inspection

The inspection process

In new installations inspection should be carried out progressively as the installation is installed, and must be done before it is energised. As far as reasonably practicable an initial inspection should be carried out to verify:

- all equipment and material is of the correct type and complies with applicable British Standards or acceptable equivalents

- all parts of the fixed installation are correctly selected and erected

- no part of the fixed installation is visibly damaged or otherwise defective

- the equipment and material used are suitable for the installation relative to the environmental conditions.

The following items must be covered in an inspection:

1. **Connection of conductors**: Every connection between conductors and equipment or other conductors shall provide durable electrical continuity and adequate mechanical strength. Requirements for the enclosure and accessibility of connections must be considered.

2. **Identification of each conductor of non-flexible cables**: Table 51A of BS 7671 provides a schedule of colour identification of each core of non-flexible cable and its conductors. It should be checked that each core of a cable is identified as necessary. Where it is desired to indicate a phased rotation or a different function for cores of the same colour, numbered sleeves are permitted.

3. **Identification of flexible cable and cord**: Protective conductors only shall be identified by a combination of green and yellow colours. Note in particular the identification in Table 51B of BS 7671 for flexible cable with up to five cores.

4. **Routing of cables**: Cable routes shall be selected with regard to the cable's suitability for the environment, i.e. ambient temperature, heat, water, foreign bodies, corrosion, impact, vibration, flora, fauna, radiation, building use and structure.

5. **Current carrying capacity**: Where practicable, the cable size should be assessed against the protective device based upon information provided by the installation designer. Reference should be made as appropriate to Appendix 4 of BS 7671.

6. **Verification of polarity**: It must be checked that no single pole switch or protective device is installed in any neutral conductor.

7. **Accessories and equipment**: Correct connection is to be checked. Table 55A of BS 7671 is a schedule of types of plug and socket outlets available, the rating and the associated British Standards. Particular attention should be paid to the requirements for a cable coupler. Lamp holders should comply with BS 5042 and be of temperature rating T2.

8. **Selection and erection to minimise the spread of fire**: A fire barrier or protection against thermal effects should be provided if necessary to meet the requirements

of BS 7671. The Regulations require that each ceiling arrangement be inspected to verify that it conforms with the manufacturer's erection instructions. This may be impossible without dismantling the system and it is essential, therefore, that inspection should be carried out at the appropriate stage of the work and that this is recorded at the time for incorporation in the inspection and test documents.

9. **Protection against direct contact:** Direct contact as defined in BS 7671 is the contact of persons or livestock with live parts. Live parts are conductors or conductive parts intended to be energised in normal use including a neutral conductor but by convention not a combined Protective Earthed Neutral (PEN) conductor. Protection is provided using the following methods:

   - Insulation. Is the insulation damaged or has too much been removed? Although protection by insulation is the usual method there are other methods of protection against direct contact.

   - Barriers. Where live parts are protected by barriers or enclosures, these should be checked for adequacy and security. Have all covers, lids and plates been securely fitted?

   - Obstacles. Protection by obstacles provides protection only against an intentional contact. If this method is used, the area shall be accessible only to skilled persons or to instructed persons under supervision. Obstacles can include a fence around a transformer sub-station and barbed wire fencing on power pylons.

   - Out of reach. Placing out of reach protects against direct contact. Increased distance is necessary where bulky conducting objects are likely to be handled in the vicinity.

10. **Protection against indirect contact**: Indirect contact as defined by BS 7671 is the contact of persons or livestock with exposed conductive parts, which have become live under fault conditions. An exposed conductive part is a conductive part of equipment which can be touched but is not live but can become live under fault conditions. Examples of exposed conductive parts could include metal trunking, metal conduit and the metal case of an electrical appliance e.g. a classroom overhead projector. Earthing provides protection against this type of fault. We also need to check that extraneous conductive parts have been correctly bonded with protective conductors. An extraneous conductive part is a conductive part that is liable to introduce a potential, generally earth potential, and not form part of an electrical installation; examples of extraneous conductive parts are metal sink tops, metal water pipes etc. The purpose of the bonding is to ensure that all extraneous conductive parts, which are simultaneously accessible, are at the same potential.

11. **Protective devices**: Have they been set correctly for the load? If rewireable fuses have been fitted, has the correct size of fuse wire been used? If a socket is to be provided for outdoor equipment, has a 30 mA rated RCD been fitted?

12. **Checks on documentation**: Diagrams, schedules, charts, instructions and any other further information must be available if inspection and testing is to be carried out in a satisfactory manner.

13. **Checks on warning notices**: These should be fixed to equipment operating in excess of 250 volts where this voltage would not normally be expected.

Periodic inspection

As previously mentioned, all installations should be regularly inspected and, as necessary, tested. In some cases there is a statutory requirement for inspections at specified intervals, but in others the period is left to the discretion of the designer, installer or the inspector.

Where a diagram, chart or tables are not available a degree of exploratory work may be necessary so that inspection and testing can be carried out safely and effectively. Note should be made of any known changes in environmental conditions, building structure and alterations, which may have affected the suitability of the wiring for its present load and method of installation.

A careful check should be made of the type of equipment on site so that the necessary precautions can be taken, where conditions permit, to disconnect or short out electronic and other equipment, which may be damaged by subsequent testing. Special care must be taken where control and protective devices contain electronic components. It is essential to determine the degree of these disconnections before planning the detailed inspection and testing. If inspection and testing cannot be carried out safely without diagrams or equivalent information, they must be prepared for compliance with Section 6 of the Health and Safety at Work Act 1974.

For safety, it is necessary to carry out a visual inspection of the installation before beginning any tests or opening enclosures, removing covers, etc. So far as is reasonably practicable, the visual inspection must verify that the safety of persons, livestock and property is not endangered. A thorough visual inspection should be made of all electrical equipment that is not concealed and should include the accessible internal condition of a sample of the equipment. External conditions should be noted and damage identified or, if the degree of protection has been impaired, the matter should be recorded on the schedule of the report. This inspection should be carried out without power supplies to the installation, wherever possible, in accordance with the *Electricity at Work Regulations 1989*.

The inspection should include a check on the condition of all electrical equipment and materials, taking into account any available manufacturer's information with regard to the following:

- safety
- wear and tear
- corrosion
- damage
- excessive loading (overloading)
- age
- external influences
- suitability.

> **Remember**
>
> With periodic inspection and testing, inspection is the vital initial operation and testing is subsequently carried out in support of that inspection

The assessment of condition should take account of known changes in conditions influencing and affecting electrical safety, e.g. extraneous conductive parts, plumbing, structural changes, etc. It would not be practicable to inspect all parts of an installation; thus a random sample should be inspected. This should include:

- Check that joints and connections are properly secured and that there is no sign of overheating.

- Check switches for satisfactory electrical and mechanical conditions.

- Check that protective devices are of the correct rating and type; check for accessibility and damage.

- Check that conductors have suffered no mechanical damage and have no signs of overheating.

- Check that the condition of enclosures remains satisfactory for the type of protection required.

Initial verification, certification and commissioning

Inspection and testing of electrical installations

This section is intended to act as a guide for those engaged in the inspection and testing of electrical installations both at the initial construction stage or when carrying out a periodic inspection and test as required by the *IEE Wiring Regulations* (BS 7671).

To ensure that this most important aspect of an electrician's work is carried out satisfactorily, the inspection and test procedure must be carefully planned and carried out and the results correctly documented. It is hoped that the principles described here will enable that goal to be achieved with the minimum of errors.

The process of commissioning is one that combines a full inspection of the installation and all tests required to be carried out before the installation is energised (referred to as pre-commissioning tests), followed by those tests which require power to be available i.e. the measurement of earth-fault loop impedance and the functional testing of residual current devices.

Commissioning also involves the initial energising of an installation, which has to be carried out in a controlled manner and with the knowledge of everyone involved. Other persons who may be working on the site at the time of energising must be informed of the likelihood of power being applied to the installation in order that all precautions can be taken to prevent danger.

Did you know?

The final act of the commissioning process is to ensure the safe and correct operation of all circuits and equipment which have been installed, and that the customer's requirements have been met. All of these points will be discussed in detail later

Initial verification – general requirements

BS 7671 Part 7 states that every electrical installation shall, either during construction and/or on completion, but certainly before being put into service, be inspected and tested to verify, so far as is reasonably practicable, that the requirements of the Regulations have been met.

When carrying out such inspection and test procedures, precautions must be taken to ensure no danger is caused to any person or livestock and to avoid damage to property and installed equipment.

Periodic inspection and test procedures must also be carried out on existing installations to ensure that the installation has not deteriorated and still meets all regulatory requirements. Details of the recommended periods between such tests are given later in this chapter.

Inspection and testing is also required where minor alterations or additions have been carried out to existing installations. In such cases care has to be taken to ensure that the original design has not been changed in such a way that a dangerous situation might occur e.g. overloading of existing cables or switchgear.

Initial verification is intended to confirm that the installation complies with the designer's requirements and has been constructed, inspected and tested in accordance with BS 7671. Inspection is a very important part of this procedure and should be carried out prior to electrical tests being applied and normally with that part of the installation under inspection disconnected from the supply.

The results of all such inspections and tests must be recorded and compared with the relevant design criteria. The relevant criteria will for the most part be the requirements of the Regulations, although there may be some instances where the designer has specified requirements that are particular to the installation concerned. In these cases the person carrying out the inspection and test should be provided with the necessary data for comparison purposes. In the absence of such data the inspector should apply the requirements set out in BS 7671.

Certification

Although the different forms of certification available will be discussed in some detail later in this chapter it is worth pointing out at this stage that BS 7671 provides examples of three different types of certificate:

- The Electrical Installation Certificate is designed for new installations or major alterations and is available in two different forms. One is for use where the responsibility for design, construction and inspection and test is the responsibility of one person and the second is for use where design, construction and inspection and test have been carried out by, and require the signatures of, three different parties.

- Minor Works Certificates should only be used for small works that do not include the provision of a new circuit, such as an additional socket outlet or lighting point to an existing circuit.

- Periodic Inspection and Test Certificates are for use where periodic inspections and tests have been carried out on existing installations and would normally have both an Inspection Schedule and a Test Results Schedule attached.

(In addition to the above, a standard form of Inspection Schedule and a Test Results Schedule are provided for use as attachments to the appropriate certificate.)

On confirmation that the results of the inspection and test comply with the relevant criteria the competent person, who may be the person carrying out the inspection and testing, should sign the inspection and test section of the appropriate certificate. If that person has also been responsible for the design and construction of the installation then they should make use of the combined certificate.

Alterations and additions

Where an installation has been the subject of an alteration or addition, then the existing installation must be inspected and tested so far as is necessary to ensure the safety of the alteration and addition. This should include as a minimum the continuity of the circuit protective conductor and earth fault loop impedance.

Safety

The testing of electrical installations can cause some degree of danger and it is the responsibility of the person carrying out the tests to ensure the safety of themselves and others. *Health and Safety Executive Guidance Note GS38 (Electrical test equipment for use by electricians)* details relevant safety procedures and should be observed in full.

When using test instruments the following points will help to achieve a safe working environment:

- The person carrying out the tests must have a thorough understanding of the equipment being used and its rating.

- The person carrying out the tests must ensure that all safety procedures are being followed e.g. erection of warning notices and barriers where appropriate.

- The person carrying out the tests should ensure that all instruments being used conform to the appropriate British Standard i.e. BS EN 61010 or older instruments manufactured to BS 5458 provided these are in good condition and have been recently calibrated.

- The person carrying out the tests should check that test leads including probes and clips are in good condition, are clean and have no cracked or broken insulation. Where appropriate the requirements of GS38 should be observed, including the use of fused test leads.

A further consideration in relation to safety when carrying out inspection and test procedures is the competence of the inspector. All persons called upon to carry out an inspection and test of an installation must be skilled and experienced and have sufficient knowledge of the type of installation to be inspected and tested to ensure no risk of injury to persons or livestock or damage to property.

Information

BS 7671 requires that the following information be provided to the person carrying out the inspection and test of an installation:

- the maximum demand of the installation expressed in amperes per phase

- the number and type of live conductors at the point of supply

- the type of earthing arrangements used by the installation, including details of equipotential bonding arrangements

- the type and composition of circuits, including points of utilisation, number and size of conductors and types of cable installed (this should also include details of the 'reference installation method' used)

- the location and description of protective devices (fuses/circuit breakers etc.)

- details of the method selected to prevent danger from shock in the event of an earth fault e.g. earthed equipotential bonding and automatic disconnection of supply

- the presence of any sensitive electronic devices.

Safety tip

Particular attention should be paid when using instruments capable of generating a test voltage in excess of 50 volts e.g. Insulation Resistance Testers. If the live terminals of such an instrument are touched a shock will be received and, although this may not be harmful in itself, it may cause a loss of concentration that could be dangerous, especially if working at heights. Care must also be taken when working with instruments that use the supply voltage for the purpose of the test, such as earth loop impedance testing or when testing a residual current device (RCD). Either of these tests can impose a voltage on associated-earthed metalwork, and precautions must be taken to avoid the risk of shock to other persons

It is a requirement of the Regulations that this information shall be available as an on-site record of the design. Such details would normally be included in the project Health and Safety File required by the Construction (Design and Management) Regulations.

The customer

Following initial verification, an Electrical Installation Certificate together with a schedule of test results shall be given to the person ordering the work. The requirements of the Regulations have not been met until this has been done. Sometimes the person ordering the work is not the end-user, e.g. the builder of a new housing estate sells the individual houses to various occupiers. In these cases it is recommended that copies of the Inspection and Test Certificates together with a Test Results Schedule be passed on to the new owners.

Statutory and non-statutory requirements

For domestic electrical installations compliance with BS 7671, which is a non-statutory document, is the only requirement. For commercial or industrial installations the requirements of the *Electricity Supply Regulations 1988* and the *Electricity at Work Regulations 1989*, both of which are statutory instruments, should also be taken into account.

Compliance with BS 7671 will in most cases satisfy the requirements of statutory Regulations such as the *Electricity at Work Regulations* but this cannot be guaranteed. It is essential to establish which statutory and other Regulations apply, and to carry out the design, the construction and the inspection and testing accordingly.

Commissioning

Prior to commencing work on a new electrical installation it is essential that a full specification is prepared. The specification should set out the detailed design and provide sufficient information to enable competent persons to construct the installation and to commission it.

Precise details of all equipment installed should be obtained from the manufacturers or suppliers in order to check that the required standards have been met, to ensure satisfactory methods of installation have been used and to provide the information necessary to confirm its correct operation.

All the above information must be included in the Operation and Maintenance Manual prepared for the project, which should also contain a description of how the installation is to operate and technical data for all items installed such as switchgear, luminaires and any special control systems that may have been incorporated.

The Health and Safety at Work Act 1974 and the *Construction (Design and Management) Regulations 1994* both state the requirements for the provision of such information. The purpose of commissioning is to ensure the safe energising and operation of all aspects of the installation and consists of the following activities:

- inspection of the installation for compliance with BS 7671, the project specification and other relevant standards

- pre-commissioning tests such as continuity, polarity and insulation resistance (including that of any fixed equipment)

- energising of individual circuits within the installation

- measurement of earth loop impedance and functional testing of any residual current devices

- checking the correct operation of all current using equipment and associated control systems

- handover of the installation to the client.

Inspection and testing procedures are fully covered in other sections. However, the requirements for energising circuits and commissioning of equipment are discussed below.

Before energising any circuit it must be assured that it is safe to do so. All required inspection and testing procedures must have been followed and the results confirmed as satisfactory. All other persons in the building e.g. other contractors, users of the building, members of the public etc. must be advised of any potential danger, and warning notices or barriers must be erected where required. All devices used to control current using equipment must be in the 'off' position.

After energising the installation, measurement of earth loop impedance and the functional testing of any RCDs must be completed and the results confirmed as satisfactory before any further commissioning is carried out.

Commissioning of individual items of equipment may now take place. This includes ensuring the correct settings of overload devices, thermostats, time controllers etc. and finally energising the equipment being commissioned in order to check its correct operation. It may be necessary where a piece of equipment requires more than one service to combine the electrical commissioning with that of other contractors e.g. a gas fired central heating system requires both gas and water as well as an electricity supply, and commissioning of the system should preferably take place in conjunction with the heating engineer.

Handover to client

Handover of the installation to the client is the final activity and should include a tour of the installation, an explanation of any specific controls or settings and, where necessary, a demonstration of any particularly complicated control systems. The Operation and Maintenance Manuals produced for the project should be formally handed to the client at this stage.

Remember

Many industrial processes have very complicated control systems and may require the supplier of the control panel as well as the client to be present when commissioning is carried out

The sequence of initial tests

BS 7671 Regulation 711-01-01 states 'Every installation shall, during erection and/or on completion before being put into service, be inspected and tested to verify, so far as is reasonably practicable, that the requirements of the Regulations have been met. Precautions shall be taken to avoid danger to persons, livestock, and to avoid damage to property and installed equipment during inspection and testing'.

Regulation 713 lists the sequence in which tests should be carried out. If any test indicates a failure to comply, that test and any preceding test, the results of which may have been influenced by the fault indicated, must be repeated after the fault has being rectified.

The sequence of tests

Initial tests should be carried out in the following sequence where applicable, before the supply is connected or with the supply disconnected as appropriate:

1. Continuity of circuit protective conductors (cpcs) including main and supplementary equipotential bonding conductors.
2. Continuity of ring final circuit conductors.
3. Insulation resistance.
4. Site applied insulation.
5. Protection by separation of circuits.
6. Protection by barriers or enclosures provided during erection.
7. Insulation of non-conducting floors and walls.
8. Polarity.
9. Earth electrode resistance.

With the electrical supply connected, re-test polarity before further testing:

10. Prospective fault current.
11. Earth-fault loop impedance.
12. Operation of residual current devices (RCDs).

The test results should be recorded on an installation schedule similar to the diagram in Figure 5.1.

Forms of completion or periodic inspection, inspection, test and an installation schedule (including test results) should be provided to the person ordering the work.

These tests are fully described in *Guidance Note 3 to BS 7671*.

Contractor			Address/location of dist board			Type of supply			Instruments		
Test date						Ze at origin			RCD tester		
			Equipment vulnerable to testing						Continuity		
Signature						PSSC			Insulation		
									Others		

		overcurrent device			circuit conductors		Test results											
Circuit	No. of points	type	rating	Short Circuit Cap	live	CPC	Continuity		Insulation resistance					Earth	Functional testing			polarity	remarks
							R1+R2	R2	Phase/ Phase	Phase/ neutral	Phase/ earth	Neutral/ earth	loop impedance	RCD	Time ms	other		
1	2	3	4	5	6	7	8	9	10	11	12	13	14	15	16	17	18	19
1																		
2																		
3																		
4																		
5																		
6																		
7																		
8																		
9																		
10																		
11																		
12																		

Main bonding check	size		Gas	
Gas				
Water			Earth electrode resistance	
Other				

Figure 5.1   Installation schedule

Continuity of circuit protective conductors (cpcs) including main and supplementary equipotential bonding conductors

Regulations state that every protective conductor, including each bonding conductor, shall be tested to verify that it is electrically sound and correctly connected. The test described below as well as checking the continuity of the protective conductor will also measure $R_1 + R_2$ which, when corrected for temperature, will enable the designer to verify the calculated earth fault loop impedance Z_s. For this test you need a low reading ohmmeter.

N.B. Neutral omitted for clarity

Figure 5.2   Test method 1

Test method 1. Before carrying out this test (shown in Figure 5.2) the leads should be 'nulled out'. If the test instrument does not have this facility, the resistance of the leads should be measured and deducted from the readings. The phase conductor and the protective conductor are linked together at the consumer unit or distribution board. The ohmmeter is used to test between the phase and earth terminals at each outlet in the circuit. The measurement at the circuit's extremity should be recorded and is the value of $R_1 + R_2$ for the circuit under test. On a lighting circuit the value of R_1 should include the switch wire at the luminaires. This method should be carried out before any supplementary bonds are made.

green/yellow

Long 'wander' lead

Low reading ohmmeter

Socket outlet

E

N   P

C.P.C.   P

C.C.U.

N.B. Neutral and phase omitted for clarity

Figure 5.3 Test method 2

Test method 2. One lead of the continuity tester is connected to the consumer's main earth terminals (see Figure 5.3). The other lead is connected to a trailing lead, which is used to make contact with protective conductors at light fittings, switches, spur outlets etc. The resistance of the test leads will be included in the result; therefore the resistance of the test leads must be measured and subtracted from the reading obtained if the instrument does not have a nulling facility. In this method the protective conductor only is tested and this reading R_2 is recorded on the installation schedule.

Test of the continuity of supplementary bonding conductors

Test method 2 as described above is used for this purpose. The ohmmeter leads are connected between the points being tested, between simultaneously accessible extraneous conductive parts i.e. pipe work, sinks etc. or between simultaneously accessible extraneous conductive parts and exposed conductive parts (metal parts of the installation). This test will verify that the conductor is sound. To check this, move the probe to the metalwork to be protected as shown in Figure 5.4. This method is also used to test the main equipotential bonding conductors.

Where ferrous enclosures have been used as the protective conductors, e.g. conduit, trunking, steel-wire armouring etc., the following special precautions should be followed:

- Perform the standard ohmmeter test using the appropriate test method described above. Use a low resistance ohmmeter for this test.

- Inspect the enclosure along its length to verify its integrity.

- If there is any doubt as to the soundness of this conductor a further test should be performed using a phase-earth loop impedance tester after the connection of the supply.

- If there is still doubt, a further test may be carried out using a high-current, low-impedance ohmmeter. The high-current, low-impedance ohmmeter has a test voltage not exceeding 50 volts and can provide a current approaching 1.5 times the design current of the circuit, but the current need not exceed 25 A.

Metal sink

Supplementary bonding conductor

Low reading ohmmeter

Metal waste pipe

M.E.T.

Figure 5.4 Test of the continuity of supplementary bonding conductors

Continuity of ring final circuit conductors

A test is required to verify the continuity of each conductor including the circuit protective conductor (cpc) of every ring final circuit. The test results should establish that the ring is complete and has no interconnections. The test will also establish that the ring is not broken. Figure 5.5 shows a ring circuit illustrating these faults.

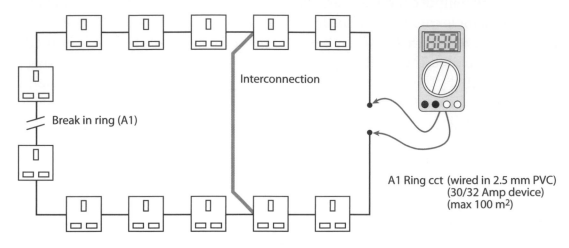

Figure 5.5 Test of continuity of ring final circuit conductors

It may be possible, as an alternative and in order to establish that no interconnected multiple loops have been made in the ring circuit, for the inspector to check visually each conductor throughout its entire length. However, in most circumstances this will not be practicable and the following test method for checking ring circuit continuity is recommended.

The phase, neutral and protective conductors are identified and their resistances are measured separately (see Figure 5.6). The end-to-end resistance of the protective conductor is divided by four to give R_2 and is recorded on the installation schedule form. A finite reading confirms that there is no open circuit on the ring conductors under test.

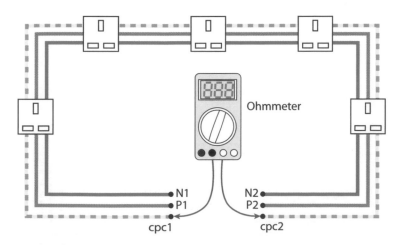

Figure 5.6 Measurement of phase, neutral and protective conductors

The resistance values obtained should be the same (within 0.05 ohms) if the conductors are the same size. If the protective conductor has a reduced csa, the resistance of the protective loop will be proportionally higher than that of the phase or neutral loop. If these relationships are not achieved then either the conductors are incorrectly identified or there is a loose connection at one of the accessories.

The phase and neutral conductors are then connected together so that the outgoing phase conductor is connected to the returning neutral conductor and vice versa as in Figure 5.7. The resistance between phase and neutral conductors is then measured at each socket outlet. The readings obtained from those sockets wired into the ring will be substantially the same and the value will be approximately half the resistance of the phase or the neutral loop resistance. Any sockets wired as spurs will have a proportionally higher resistance value corresponding to the length of the spur cable.

Figure 5.7   Phase and neutral conductors connected together

The latter exercise is then repeated but with the phase and cpc cross-connected as in Figure 5.8. The resistance between phase and earth is then measured at each socket. The highest value recorded represents the maximum $R_1 + R_2$ of the circuit and is recorded on the test schedule. It can be used to determine the earth loop impedance (Z_s) of the circuit to verify compliance with the loop impedance requirements of the Regulations.

Figure 5.8   Phase and cpc cross-connected

Insulation resistance

Insulation resistance tests are to verify, for compliance with BS 7671, that the insulation of conductors, electrical accessories and equipment is satisfactory and that electrical conductors and protective conductors are not short-circuited, or do not show a low insulation resistance (which would indicate defective insulation). In other words, we are testing to see whether the insulation of a conductor is so poor as to allow any conductor to 'leak' to earth or to another conductor. Before testing ensure that:

- Pilot or indicator lamps and capacitors are disconnected from circuits to avoid an inaccurate test value being obtained.

- Voltage sensitive electronic equipment such as dimmer switches, delay timers, power controllers, electronic starters for fluorescent lamps, emergency lighting, Residual Current Devices etc. are disconnected so that they are not subjected to the test voltage.

- There is no electrical connection between any phase and neutral conductor (e.g. lamps left in).

To illustrate why we remove lamps, consider Figure 5.9. Should we leave the lamp in? The coil that is the lamp filament is effectively creating a short circuit between the phase and neutral conductors.

Figure 5.9 Insulation resistance test

The test equipment to be used would be an insulation resistance tester meeting the criteria as laid down in BS 7671, with insulation resistance tests carried out using the appropriate d.c. test voltage as specified in Table 71A of BS 7671 (shown here as Table 5.2). The installation will be deemed to conform with the Regulations if, with the main switchboard and each distribution circuit tested separately, with all its final circuits connected but with current using equipment disconnected, it has an insulation resistance not less than that specified in Table 71A.

Table 71A Minimum value of insulation resistance		
Circuit nominal voltage (V)	Test voltage dc (V)	Minimum insulation resistance
SELV and PELV	250	0.25
Up to and including 500 V with the exception of the above systems	500	0.5
Above 500 V	1000	1.0

Table 5.2   Minimum value of insulation resistance (BS 7671)

The tests should be carried out with the main switch off, all fuses in place, switches and circuit breakers closed, lamps removed, and fluorescent and discharge luminaires and other equipment disconnected. Where the removal of lamps and/or the disconnection of current using equipment is impracticable, the local switches controlling such lamps and/or equipment should be open (see Regulation 713-04).

Simple installations that contain no distribution circuits should be tested as a whole; however, to perform the test in a complex installation it may need to be subdivided into its component parts. Although an insulation resistance value of not less than 0.5 megohms complies with the Regulations, where an insulation resistance value of less than 2 megohms is recorded, the possibility of a latent defect exists. If this is the case, each circuit should be separately tested and its measured insulation resistance should be greater than 2 megohms.

Test 1 – Insulation resistance between live conductors

- Single-phase circuits – Test between the phase and neutral conductors at the appropriate switchboard.

- Three-phase circuits – Before this test is carried out you must ensure that the incoming neutral has been disconnected so there is no connection with earth.

Now make a series of tests between live conductors in turn at the appropriate switchboard as follows:

- between Brown phase and to the Black phase, Grey phase and Neutral (Blue) grouped together

- between Black phase and to the Grey phase and Neutral (Blue) grouped together

- between Grey phase and Neutral (Blue).

Where it is not possible to group conductors in this way, conductors may be tested singly. Figure 5.10 illustrates the testing of insulation resistance on a single-phase lighting circuit. But remember on a two-way circuit to operate the two-way switch and re-test to make sure that you have tested all of the strappers!

Figure 5.10   Insulation resistance test on single-phase lighting circuit

Test 2 – Insulation resistance from earth to phase and neutral connected together

- Single-phase circuits – Test between the phase and neutral conductors and earth at the appropriate distribution board. Where any circuits contain any two-way switching, the two-way switches will require to be operated and another insulation resistance test carried out, including the two-way strapping wire which was not previously included in the test.

- Three-phase circuits – Measure between all phase conductors and neutral bunched together, and earth. Where a low reading is obtained (less than 2 megohms) it may be necessary to test each conductor to earth separately. Figure 5.11 shows the test of insulation resistance between earth to phase and neutral connected together on a socket outlet circuit.

Figure 5.11   Insulation resistance test between earth to phase and neutral connected together on socket outlets

Some electricians prefer, or find it quicker, to test between individual conductors rather than to group them together. As an example, Figure 5.12 shows the ten readings that would need to be taken for a three-phase circuit.

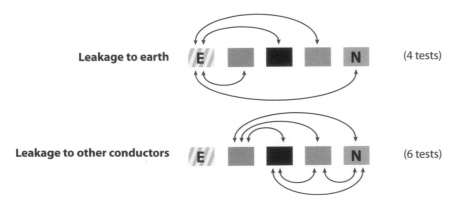

Figure 5.12   Three-phase circuit test

Polarity

A test needs to be performed to check the polarity of all circuits. This must be done before connection to the supply, with either an ohmmeter or the continuity range of an insulation and continuity tester.

The purpose of the test is to confirm that all protective devices and single-pole switches are connected to the phase conductor and that the phase terminal in socket outlets and the centre contact of screw-type lamp holders are also connected to the phase conductor. (Also a check should be made to ensure that the polarity of the incoming supply is correct, otherwise the whole installation would have the wrong polarity.)

In essence, having established the continuity of the cpc in an earlier test, we now use this as a long test lead, temporarily linking it out with the circuit phase conductor at the distribution board and then making our test across the phase and earth terminals at each item in the circuit under test. Remember to close lighting switches before carrying out the test.

Figure 5.13 shows the test for polarity of a lighting circuit using a continuity tester.

For ring circuits, if the tests required by Regulation 713-03 ring circuit continuity have been carried out, the correct connections of phase, neutral and cpc conductors will have been verified and no further testing is required. For radial circuits the $R_1 + R_2$ measurements should also be made at each point, using this method.

Figure 5.13 Polarity test of lighting circuit using continuity tester

Earth electrode resistance (Regulation 713-10)

Where the earthing system incorporates an earth electrode as part of the system, the electrode resistance to earth needs to be measured. In previous years, the metal pipes of water mains were used, but the change to more plastic pipe work means this practice can no longer be relied upon.

Some of the types of accepted earth electrode are:

- earth rods or pipes
- earth tapes or wires
- earth plates
- underground structural metalwork embedded in foundations
- lead sheaths or other metallic coverings of cables
- metal pipes.

The resistance to earth depends upon the size and type of electrode used, and remember that we want as good a connection to earth as possible. The connection to the electrode must be made above ground level.

Measurement by standard method

When measuring earth electrode resistances to earth where low values are required, as in the earthing of the neutral point of a transformer or generator, test method 1 below may be used, using an earth electrode resistance tester.

Test method 1

Before this test is undertaken, the earthing conductor to the earth electrode must be disconnected either at the electrode or at the main earthing terminal. This will ensure that all the test current passes through the earth electrode alone. However, as this will leave the installation unprotected against earth faults, switch off the supply before disconnecting the earth.

The test should be carried out when the ground conditions are at their least favourable, i.e. during a period of dry weather, as this will produce the highest resistance value.

The test requires the use of two temporary test electrodes (spikes) and is carried out in the following manner:

- Connect the earth electrode to terminals C_1 and P_1 of a four-terminal earth tester. To exclude the resistance of these test leads from the resistance reading, individual leads should be taken from these terminals and connected separately to the electrode. However, if the test lead resistance is insignificant, the two terminals may be short-circuited at the tester and connection made with a single test lead, the same being true if you are using a three-terminal tester.

- Connection to the temporary 'spikes' is now made as shown in Figure 5.14. The distance between the test spikes is important. If they are too close together, their resistance areas will overlap. In general, you can expect a reliable result if the distance between the electrode under test and the current spike is at least ten times the maximum length of the electrode under test, e.g. 30 m away from a 3 m long rod electrode. (Resistance area is important where livestock are concerned, as the front legs of an animal may be outside, and the back legs of the same animal inside the resistance area, thus creating a potential difference. As little as 25 V can be lethal to them, so it's important to ensure that all of the electrode is well below ground level and RCD protection is used).

E = the electrode under test
C_2 = a temporary test spike/electrode
P_2 = a temporary test spike/electrode

Figure 5.14   Earth electrode test

- We then take three readings:

  1. first with the potential spike initially midway between the electrode and current spike

  2. second at a position 10 per cent of the electrode-to-current spike distance back towards the electrode

  3. third at a position 10 per cent of the distance towards the current spike.

- By comparing the three readings, a percentage deviation can be determined. We do this calculation by taking the average of the three readings, finding the maximum deviation of the readings from this average in ohms, and expressing this as a percentage of the average. The accuracy of the measurement using this technique is on average about 1.2 times the percentage deviation of the readings. It is difficult to achieve an accuracy better than 2 per cent, and you should not accept readings that differ by more than 5 per cent. To improve the accuracy of the measurement to acceptable levels, the test must be repeated with larger separation between the electrode and the current spike.

Once the test is completed, make sure that the earthing conductor is re-connected.

Test method 2

Guidance Note 3 to BS 7671 lists a Test method 2. However, this is an alternative method for use on RCD protective TT installations only.

Earth-fault loop impedance

When designing an installation, it is the designer's responsibility to ensure that, should a phase-to-earth fault develop, the protection device will operate safely and within the time specified by BS 7671. Although the designer can calculate this in theory, it is not until the installation is complete that the calculations can be checked.

It is necessary therefore to determine the earth-fault loop impedance (Z_s) at the furthest point in each circuit and to compare the readings obtained with either the designer's calculated values or the values tabulated in BS 7671 or the *IEE On-Site Guide*.

The earth-fault loop is made up of the following elements:

- the phase conductor from the source of the supply to the point of the fault

- the circuit protective conductor

- the main earthing terminal and earthing conductor

- the earth return path (dependent on the nature of the supply, TN-S, TN-C-S etc.)

- the path through the earthed neutral of the supply transformer

- the secondary winding of the supply transformer.

Figure 5.15 Earth-fault loop

The value of earth-fault loop impedance may be determined by:

- direct measurement of Z_s

- direct measurement of Z_e at the origin of the circuit and adding to this the value of $R_1 + R_2$ measured during continuity tests: $Z_s = Z_e + (R_1 + R_2)$

- obtaining the value of Z_e from the electricity supplier and adding to this the value of $R_1 + R_2$ as above. However, where the value of Z_e is obtained from the electricity supplier and is not actually measured, a test must be carried out to ensure that the main earthing terminal is in fact connected to earth using an earth loop impedance tester or an approved test lamp.

Direct measurement of Z_s

Direct measurement of earth-fault loop impedance is achieved by use of an earth-fault loop impedance tester, which is an instrument designed specifically for this purpose. The instrument operates from the mains supply and therefore can only be used on a live installation.

The instrument is usually fitted with a standard 13 A plug for connecting to the installation directly through a normal socket outlet, although test leads and probes are also provided for taking measurements at other points on the installation.

In order to eliminate any parallel earth return paths, the main equipotential bonding conductors are disconnected for the duration of the test. This will ensure that the reading is not distorted by the presence of gas or water service pipes acting as part of the earth return path. Precautions must be taken, however, to ensure that the main equipotential bonding conductors are reconnected after the test has taken place.

Earth-fault loop impedance testers are connected directly to the circuit being tested and care must be taken to prevent danger. Should a break have occurred anywhere in the protective conductor under test then the whole of the earthing system could become live. It is essential therefore that protective conductor continuity tests be carried out prior to the testing of earth-fault loop impedance. Communication with other users of the building and the use of warning notices and barriers is essential.

Measurement of Z_e

The value of Z_e can be measured using an earth fault loop impedance tester at the origin of the installation. However, as this requires the removal of covers and the exposure of live parts, extreme care must be taken and the operation must be supervised at all times. The instrument is connected using approved leads and probes between the phase terminal of the supply and the means of earthing with the main switch open or with all sub-circuits isolated. In order to remove the possibility of parallel paths, the means of earthing must be disconnected from the equipotential bonding conductors for the duration of the test. With the instrument correctly connected and the test button pressed, the instrument will give a direct reading of the value of Z_e. It must be remembered to re-connect all earthing connections on completion of the test.

Verification of test results

The values of earth loop impedance obtained (Z_s) should be compared with one of the following:

- for circuits wired in standard PVC insulated cables, the values given in the *IEE On-Site Guide* after correction for ambient temperature where necessary
- the earth-fault loop impedance values determined by the designer
- the values given in Tables 41B1, 41B2 or 41D of BS 7671 *IEE Wiring Regulations* after being corrected for the difference in temperature between normal conductor operating temperature and the ambient temperature when the test takes place
- rule of thumb.

Standard PVC circuits

The measured loop impedances given in the *IEE On-Site Guide* (repeated in Tables 5.3 to 5.7 overleaf) have been determined to ensure compliance with the required disconnection times where conventional final circuits are installed. The values assume an ambient temperature when carrying out the tests of between 10–20°C. At ambient temperatures other than this, correction factors shown in Table 5.6 should be applied.

Design values

Where the designer of the installation provides calculated values of earth-loop impedance, the measured values should be compared with these.

Tables 41B1, 41B2 and 41D

You may need to refer to BS 7671 2001 for measured values of earth loop impedance given in Tables 41B1, 41B2 and 41D of BS 7671, which should not be exceeded when the conductors are at their normal operating temperature. If the conductors are at a different temperature when tested, the reading should be adjusted accordingly.

Rule of thumb

As an approximation or 'rule of thumb' the measured value of earth-fault loop impedance (Z_s) at the most remote outlet should not exceed three-quarters of the relevant value given in Tables 41B1, 41B2 or 41D.

Warning: If the circuit being tested is protected by an RCD, the test procedure may cause the RCD to operate causing unwanted isolation of the circuit. Certain types of test instrument may be used that are specifically designed to overcome this problem; otherwise it will be necessary to measure the value of R_1 and R_2 with the circuit isolated and add this to the value of Z_e measured at the incoming terminals.

(a) 0.4 second disconnection time   NP = not permitted

Protective Conductor sq mm	Fuse Rating (amps)				
	5	15	20	30	45
1.0	8.00	2.14	1.48	NP	NP
1.5	8.00	2.14	1.48	0.91	NP
2.5 to 16.0	8.00	2.14	1.48	0.91	0.5

(b) 5 second disconnection time   NP = not permitted

Protective Conductor sq mm	Fuse Rating (amps)				
	5	15	20	30	45
1.0	14.8	4.46	2.79	NP	NP
1.5	14.8	4.46	3.20	2.08	NP
2.5	14.8	4.46	3.20	2.21	1.20
4.0 to 16.0	14.8	4.46	3.20	2.21	1.33

Table 5.3   *IEE On-Site Guide* Table 2A. Maximum measured value of earth loop impedance (in ohms) when the overcurrent protective device is a fuse to BS 3036

(a) 0.4 second disconnection time   NP = not permitted

Protective Conductor sq mm	Fuse Rating (amps)							
	6	10	16	20	25	32	40	50
1.0	7.11	4.26	2.26	1.48	1.20	0.69	NP	NP
1.5	7.11	4.26	2.26	1.48	1.20	0.87	0.67	NP
2.5 to 16.0	7.11	4.26	2.26	1.48	1.20	0.87	0.69	0.51

(b) 5 second disconnection time   NP = not permitted

Protective Conductor sq mm	Fuse Rating (amps)							
	6	10	16	20	25	32	40	50
1.0	11.28	6.19	3.20	1.75	1.24	0.69	NP	NP
1.5	11.28	6.19	3.49	2.43	1.60	1.12	0.67	NP
2.5	11.28	6.19	3.49	2.43	1.92	1.52	1.13	0.56
4.0	11.28	6.19	3.49	2.43	1.92	1.52	1.13	0.81
6.0 to 16.0	11.28	6.19	3.49	2.43	1.92	1.52	1.13	0.87

Table 5.4   *IEE On-Site Guide* Table 2B. Maximum measured earth-fault loop impedance (in ohms) when the overcurrent device is a fuse to BS 88

(a) 0.4 second disconnection time NP = not permitted

Protective Conductor sq mm	Fuse Rating (amps)				
	5	15	20	30	45
1.0	8.72	2.74	1.42	0.80	NP
1.5	8.72	2.74	1.42	0.96	0.34
2.5 to 16.0	8.72	2.74	1.42	0.96	0.48

(b) 5 second disconnection time NP = not permitted

Protective Conductor sq mm	5	15	20	30	45
1.0	13.68	4.18	1.75	0.80	NP
1.5	13.68	4.18	2.34	1.20	0.34
2.5	13.68	4.18	2.34	1.54	0.53
4.0	13.68	4.18	2.34	1.54	0.70
6.0 to 16.0	13.68	4.18	2.34	1.54	0.80

Table 5.5 *IEE On-Site Guide* Table 2C. Maximum measured earth fault loop impedance (in ohms) when the overcurrent protective device is a fuse to BS 1361

Ambient temperature (degrees centigrade)	Correction factors
0	0.96
5	0.98
10	1.00
25	1.06
30	1.08

Table 5.6 Ambient temperature correction factors

MCB type	MCB rating (amps)												
	5	6	10	15	16	20	25	30	32	40	45	50	63
1	9.60	8.00	4.80	3.20	3.00	2.40	1.92	1.60	1.50	1.20	1.06	0.96	0.76
2	5.49	4.57	2.74	1.83	1.71	1.37	1.10	0.91	0.86	0.69	0.61	0.55	0.43
B	–	6.40	3.84	–	2.40	1.92	1.54	–	1.20	0.96	0.86	0.77	0.61
3 or C	3.84	3.20	1.92	1.28	1.20	0.96	0.77	0.64	0.60	0.48	0.42	0.38	0.30
D	1.92	1.60	0.96	0.64	0.60	0.48	0.38	0.32	0.30	0.24	0.22	0.19	0.15

Table 5.7 *IEE On-Site Guide* Table 2D. Maximum measured earth fault loop impedance (in ohms) when overcurrent protective device is an MCB to BS 3871 or BS EN 60898.

Operation of residual current devices (RCDs)

Although the majority of residual current devices (RCDs) in use are designed to operate at earth leakage currents not exceeding 30 mA, other values of operating current are available for use in special circumstances. All RCDs are electromechanical devices that must be checked regularly to confirm that they are still in working order. This can be done at regular intervals by simply pressing the test button on the front of the device.

Where an installation incorporates an RCD, Regulation 514-12-02 requires a notice to be fixed in a prominent position at or near the origin of the installation.

The integral test button incorporated in all RCDs only verifies the correct operation of the mechanical parts of the RCD and does not provide a means of checking the continuity of the earthing conductor, the earth electrode or the sensitivity of the device. This can only be done effectively by use of an RCD tester specifically designed for testing RCDs as described below.

Test method

The test must be made on the load side of the RCD between the phase conductor of the protected circuit and the associated circuit protective conductor. The load being supplied by the RCD should be disconnected for the duration of the test. The test instrument is usually fitted with a standard 13 A plug top, and the easiest way of making these connections, wherever possible, is by plugging the instrument into a suitable socket outlet protected by the RCD under test.

The test instrument operates by passing a simulated fault current of known value through the RCD and then measures the time taken for the device to trip. Although different types of RCD have different requirements (time delays etc.), for general purpose RCDs the test criteria are as follows:

- With an earth leakage current equivalent to 50 per cent of the rated tripping current of the RCD, the device should not open (this is to ensure that nuisance tripping due to leakage currents less than the rated tripping current will not occur).

- For general purpose RCDs to BS 4293 an earth leakage current equivalent to 100 per cent of the rated tripping current of the RCD, should cause the device to open in less than 200 milliseconds.

- Individual RCD-protected socket outlets should also meet the above criteria.

- For general purpose RCDs to BS EN 61008 or RCBOs to BS EN 61009 an earth leakage current equivalent to 100 per cent of the rated tripping current of the RCD should cause the device to open in less than 300 milliseconds unless it is of 'Type S' which incorporates an intentional time delay in which case it should trip within the time range of 130 to 500 milliseconds.

- Where an RCD with an operating current not exceeding 30 mA is used to provide supplementary protection against direct contact, a test current of five times the rated trip current (e.g. $5 \times 30 = 150$ mA) should cause the RCD to open in less than 40 milliseconds.

A further mechanical function check should be carried out to ensure that the operating switch of the residual current device is functioning correctly. Results of all the above tests should be recorded on the appropriate test results schedule.

Safety

Under certain circumstances these tests can result in potentially dangerous voltages appearing on exposed and extraneous conductive parts within the installation and therefore suitable precautions must be taken to prevent contact by persons or livestock with any such part. Other building users should be made aware of the tests being carried out and warning notices should be posted as necessary.

Electrical test instruments

BS EN 61010 covers basic safety requirements for electrical test instruments, and all instruments should be checked for conformance with this standard before use. Older instruments may have been manufactured in accordance with BS 5458 but, provided these are in good condition and have been recently calibrated, there is no reason why they cannot be used.

Instruments may be analogue (i.e. fitted with a needle that gives a direct reading on a fixed scale) or digital, where the instrument provides a numeric digital visual display of the actual measurement being taken. Insulation and continuity testers can be obtained in either format whilst earth-fault loop impedance testers and RCD testers are digital only.

Calibration and instrument accuracy

To ensure that the reading being taken is reasonably accurate, all instruments should have a basic measurement accuracy of at least 5 per cent. In the case of analogue instruments a basic accuracy of 2 per cent of full-scale deflection should ensure the required accuracy of measured values over most of the scale.

All electrical test instruments should be calibrated on a regular basis. The time between calibrations will depend on the amount of usage that the instrument receives, although this should not exceed 12 months in any circumstances. Instruments have to be calibrated in laboratory conditions against standards that can be traced back to national standards; therefore this usually means returning the instrument to a specialist test laboratory.

On being calibrated the instrument will have a calibration label attached to it stating the date the calibration took place and the date the next calibration is due (see Figure 5.16 overleaf). It will also be issued with a calibration certificate detailing the tests that have been carried out and a reference to the equipment used. The user of the instrument should always check to ensure that the instrument is within calibration before being put to use.

A further adhesive label (example shown in Figure 5.17) is often placed over the joint in the instrument casing stating that the calibration is void should the seal be broken. A broken seal will indicate whether anyone has deliberately opened the instrument and possibly tampered with the internal circuitry.

Instrument serial No. : ....

Date tested : ....

Date next due : ....

Figure 5.16   Adhesive calibration label

Calibration void if seal is broken

Figure 5.17   Calibration seal

Instruments that are subject to any electrical or mechanical misuse (e.g. if the instrument is subject to an electrical short circuit or is dropped) should be returned for re-calibration before being used again.

Electrical test instruments are relatively delicate and expensive items of equipment and should be handled in a careful manner. When not in use they should be stored in clean, dry conditions at normal room temperature. Care should also be taken of instrument leads and probes to prevent damage to their insulation and to maintain them in a good, safe working condition.

Types of instrument

Low-resistance ohmmeters

Where low resistance measurements are required when testing earth continuity, ring circuit continuity and polarity, then a low reading ohmmeter is required. This may be a specialised low-reading ohmmeter or the continuity scale of a combined insulation and continuity tester. Whichever type is used it is recommended that the test current should be derived from a source of supply not less than 4 V and no greater than 24 V with a short circuit current not less than 200 mA (instruments manufactured to BS EN 61557 will meet the above requirements).

Errors in the reading obtained can be introduced by contact resistance or by lead resistance. Although the effects of contact resistance cannot be eliminated entirely and may introduce errors of 0.01 ohm or greater, lead resistance can be eliminated either by clipping the leads together and zeroing the instrument before use, where this facility is provided, or alternatively measuring the resistance of the leads and subtracting this from the reading obtained.

Remember

When using an instrument out on site, the accuracy of the instrument will probably not be as good as the accuracy obtained under laboratory conditions. Operating accuracy is always worse than basic accuracy and can be affected by battery condition, generator cranking speed, ambient temperature, instrument alignment or loss of calibration

Remember

Records of calibrations carried out and copies of all calibration certificates issued should be retained in a safe place in case it is necessary to validate any test results carried out at some later date

Insulation resistance ohmmeters

Insulation resistance should have a high value and therefore insulation resistance meters must have the ability to measure high resistance readings. The test voltage required for measuring insulation resistance is given in BS 7671 Table 71A as shown in Table 5.8.

Circuit Nominal Voltage (volts)	Test Voltage d.c. (volts)	Minimum Insulation Resistance (megohms)
SELV and PELV	250	0.25
Up to and including 500 V with the exception of the above	500	0.5
Above 500 V	1000	1.0

Table 5.8 Test voltage required for measuring insulation resistance (from BS 7671)

Modern insulation and continuity tester

The photograph shows a typical modern insulation and continuity tester that will measure both low values of resistance for use when carrying out continuity and polarity tests and also high values of resistance when used for insulation resistance tests.

Instruments of this type are usually enclosed in a fully insulated case for safety reasons and have a range of switches to set the instrument correctly for the type of test being carried out i.e. continuity or insulation. The instrument also has a means of selecting the voltage range required e.g. 250 V, 500 V, 1000 V.

Other features of this particular type of instrument are the ability to lock the instrument in the 'on' position for hands-free operation and an automatic nulling device for taking account of the resistance of the test leads.

Earth-loop impedance testers

Earth-loop impedance testers of the type shown in the photograph have the capability to measure both earth-loop impedance and also prospective short-circuit current, depending on which function is selected on the range selection switch. The instrument also has a series of LED warning lights to indicate whether the polarity of the circuit under test is correct or not.

Earth-loop impedance tester

The instrument gives a direct digital read-out of the value of the measurement being taken at an accuracy of plus or minus 2 per cent.

RCD testers

Instruments for testing residual current devices, such as the one shown in the photograph, have two selection switches. One switch that should be set to the rated tripping current of the RCD (e.g. 30 mA, 100 mA etc.) and the other set to the test current required i.e. 50 per cent or 100 per cent of the rated tripping current or 150 mA for testing 30 mA RCDs when being used for supplementary protection.

RCD tester

A modern innovation by manufacturers is the production of an 'all in one' instrument that has the ability to carry out the most common tests required by the Regulations. These are:

- continuity tests (including polarity tests)

- insulation resistance tests

- earth-loop impedance tests

- RCD tests

- measurement of prospective short circuit current.

The photograph below shows an example of this type of instrument, which by manipulation of the function and range switches, will perform all of the above tests.

All-in-one RCD tester

Certificates

Electrical Installation Certificate

BS 7671 requires that, following the inspection and testing of all new installations, alterations and additions to existing installations or periodic inspections, an Electrical Installation Certificate together with a schedule of test results shall be given to the person ordering the work.

Examples of such forms are provided in Appendix 6 of the Regulations and are also referred to in the *IEE On-Site Guide* and *IEE Guidance Note 3*. Other professional organisations such as the Electrical Contractors Association (ECA) or the National Inspection Council for Electrical Installation Contractors (NICEIC) produce their own versions of these forms for their own members' use. The different types of forms available are as follows.

- Periodic Inspection Report
- Electrical Installation Certificate (Type 1)
- Electrical Installation Certificate (Type 2)
- Inspection Schedule
- Schedule of Test Results
- Electrical Installation – Minor Works Certificate

Periodic Inspection Report

The Periodic Inspection Report is for use when carrying out a routine periodic inspection and test of an existing installation and is not for use when alterations and additions are made. An Inspection Schedule and a Schedule of Test Results should accompany the Periodic Inspection Report. The general requirements for the content and completion of all the above documents are stated in the introduction to Appendix 6 of BS 7671 and are summarised below:

- Electrical Installation Certificates and Electrical Installation – Minor Works Certificates must be completed and signed by a competent person or persons in respect of the design, the construction, and the inspection and test of the installation.
- Periodic Inspection Reports must be completed and signed by a competent person in respect of just the inspection and test of the installation.
- Competent persons must have a sound knowledge and relevant experience of the type of work being undertaken and of the technical requirements of BS 7671. They must also have a sound knowledge of the inspection and testing procedures contained in the Regulations and must use suitable testing equipment.

- Electrical Installation Certificates must identify who is responsible for the design, construction, and inspection and testing, whether this is new work or an alteration or addition to an existing installation. As explained below this may be one person or it may be a number of different organisations.

- Minor Works Certificates must also identify who is responsible for the design, construction, and inspection and testing of the work carried out.

- Periodic Inspection Reports must indicate who is responsible for carrying out the periodic inspection and testing of the existing installation.

- A Schedule of Test Results must be issued with all Electrical Installation Certificates and Periodic Inspection Reports.

- When signing the above forms on behalf of a company or other business organisation, individuals should state for whom they are acting.

- For larger or more complex installations, additional documentation such as inspection checklists may be required for clarification purposes.

The design, construction, and inspection and testing of an installation may be the responsibility of just one person, in which case the shortened version of the Electrical Installation Certificate may be used.

On larger or more complex installations the design, construction, and inspection and test may be the responsibility of a number of different organisations, in which case all the parties concerned should be identified and must sign the certificate in the appropriate place. This may be the case where a design consultant is employed to carry out the design of the installation and an electrical contractor is employed to carry out the construction.

Many larger companies now employ dedicated inspectors whose job it is to inspect and test the work of other electricians within the company. In this case the certificate may be signed by the person carrying out the installation work as the constructor and the person carrying out the inspection and test as the inspector.

Provided that each party is clearly identified in the appropriate place, this is quite acceptable. In other cases the inspection and testing of the installation may be sub-contracted to a second contractor, in which case they should sign the certificate for that section of the work.

The certificate also provides accommodation for the identification and signatures of two different design organisations. This is for use where different sections of the installation are designed by different parties e.g. the electrical installation may be designed by one consultant and the fire-alarm system or other specialist service may be designed by another. If this should be the case then the details and signatures of both parties are required. BS 7671 provides advice notes for the completion of each type of form and also provides guidance notes for the person receiving the certificate. These are particularly helpful for non-technical persons.

Electrical Installation Certificate (Type 1)

For use when inspecting and testing a new installation, a major alteration or addition to an existing installation where the design, construction, and inspection and testing of the installation are the responsibility of different organisations. See Figure 5.18.

For Design

I/We being the person(s) responsible for the design of the Electrical Installation (as indicated by my/our signatures) particulars of which are described above, having exercised reasonable skill and care when carrying out the design hereby CERTIFY that the design work for which I/we have been responsible is to the best of my/our knowledge and belief in accordance with BS7671 amended to .... except for the departures (if any) detailed below.

Details of departures from BS7671 (if any)

....
....
....

The extent of liability of the signatory is limited to the work described above.
For the DESIGN of the installation.

Signature : .... Date : .... Name : .... Designer No.1

Signature : .... Date : .... Name : .... Designer No.2

For Construction

I/We being the person(s) responsible for the construction of the Electrical Installation (as indicated by my/our signatures) particulars of which are described above, having exercised reasonable skill and care when carrying out the construction hereby CERTIFY that the construction work for which I/we have been responsible is to the best of my/our knowledge and belief in accordance with BS7671 amended to .... Except for the departures (if any) detailed below.

Details of departures from BS7671 (if any)

....
....
....

The extent of liability of the signatory is limited to the work described above.
For the CONSTRUCTION of the Installation.

Signature : .... Date : .... Name : .... Constructor

For Inspection & testing

I/we being the person(s) responsible for the inspection and testing of the Electrical Installation (as indicated by my/our signatures) particulars of which are described above, having exercised reasonable skill and care when carrying out the inspection and testing hereby CERTIFY that the work for which I/we have been responsible is to the best of my/our knowledge and belief in accordance with BS7671 amended to .... except for the departures detailed below.

Details of departures from BS7671 (if any)

....
....
....

The extent of liability of the signatory is limited to the work described above.
For the INSPECTION & TEST of the installation

Signature : .... Date : .... Name : .... Inspector

Figure 5.18   Electrical Installation Certificate (Type 1)

As stated previously, the full certificate requires not only the signatures of those persons responsible for the design, construction, and inspection and testing of the installation but also their full details including the company they represent. See Figure 5.19.

Particulars of the Signatories to the Electrical Installation Certificate

Designer No.1

Name : .... Company : ....
Address : ....
....
Post Code : .... Tel No : ....

Designer No.2
(if applicable)

Name : .... Company : ....
Address : ....
....
Post Code : .... Tel No : ....

Constructor

Name : .... Company : ....
Address : ....
....
Post Code : .... Tel No : ....

Inspector

Name : .... Company : ....
Address : ....
....
Post Code : .... Tel No : ....

Figure 5.19   Signatories to full certificate

Electrical Installation Certificate (Type 2)

A shortened version of the Type 1 form used when inspecting and testing a new installation, major alteration or addition to an existing installation where one person is responsible for the design, construction, and inspection and testing of the installation. See Figure 5.20.

This certificate is intended for use with new installations or alterations or additions to an existing circuit where new circuits have been installed. The original certificate is to be given to the person ordering the work and a copy retained by the contractor. This certificate is not valid unless accompanied by a Schedule of Test Results.

The signatures on the certificate should be those of the persons authorised by the companies carrying out the work of design, construction, and inspection and testing.

The recommended time to the next inspection should be stated, taking into account the frequency and quality of maintenance that the installation is likely to receive.

The prospective fault current should be the greater of either that measured between live conductors or that between phase conductor(s) and earth.

Guidance Notes (for those receiving the certificate)

- This certificate has been issued to confirm that the electrical installation work to which it relates has been designed, constructed, and inspected and tested in accordance with BS 7671 (*IEE Wiring Regulations*).

- As the person ordering the work you should have received the original certificate and the contractor should have retained a copy. However, if you are not the user of the installation (the occupier) then you should pass the certificate or a full copy (including schedules) to the user immediately.

- The original certificate should be retained in a safe place to be shown to anyone carrying out further work on the installation at a later date (this includes periodic

Figure 5.20   Electrical Installation Certificate
(Type 2 Short form)

inspection). If you later vacate the property, this certificate will demonstrate that the installation did comply with BS 7671 at the time the certificate was issued. The Construction (Design and Management) (CDM) Regulations require that copies of the Installation Certificate together with any attached schedules are retained in the project Health and Safety File.

- To ensure the safety of the installation it will need to be inspected at regular intervals throughout its life. The recommended period between inspections is stated on the certificate.

- This certificate is only intended to be used for new installations or for new work such as an alteration or addition to an existing installation. For the routine inspection and testing of an existing installation a 'Periodic Inspection Report' should be issued.

- This certificate is only valid if it is accompanied by the appropriate Schedule of Test Results.

Inspection Schedule

Figure 5.21 Inspection Schedule

The Inspection Schedule provides confirmation that a visual inspection has been carried out as required by Part 7 of BS 7671 and lists all the inspection requirements of Regulation 712-01-03. The completed Inspection Schedule is attached to and forms part of the Electrical Installation Certificate.

Each item on the inspection schedule (Figure 5.21) should be checked and either ticked as satisfactory or ruled out if not applicable. In almost all cases, protection against direct contact will be by the presence of insulation and the enclosure of live parts, and protection against indirect contact will be by earthed equipotential bonding and automatic disconnection of supply (together with the presence of earthing and bonding conductors). Therefore other means of protection against either direct or indirect contact referred to below can be ruled out as not applicable.

Some example points are given below.

- SELV is an extra low-voltage system which is electrically separate from earth and from other systems. Where this method is used the particular requirements of the *IEE Wiring Regulations* must be checked (see Regulation 411-02).

- Protection against direct contact by the use of obstacles, or placing out of reach, are usually only employed in special circumstances but where they are used the requirements of the Regulations must be met (see Regulations 412-04 and 412-05).

- The use of class II equipment as a means of protection against indirect contact is infrequently adopted and only when the installation is to be supervised and under the control of a qualified electrical engineer (Regulation 413-03 applies).

- Non-conducting locations also require special precautions to be taken and are not applicable to domestic locations (see Regulation 413-04).

- Earth-free equipotential bonding as a means of protection against indirect contact is also only used in special circumstances and again is not applicable in domestic installations.

- For the requirements of systems using electrical separation as a means of protection against indirect contact, see Regulation 413-06.

On completion of the Inspection Schedule it should be signed and dated by the person responsible for carrying out the inspection.

Schedule of Test Results

The Schedule of Test Results is a written record of the results obtained when carrying out the electrical tests required by Part 7 of BS 7671. The following notes give guidance on the compilation of the Schedule, which should be attached to the Electrical Installation Certificate. See Figure 5.22.

1. The type of supply should be determined from either the supply company or by inspection of the installation.

2. Z_e should be measured with the main bonding disconnected. If Z_e is determined from information provided by the supply company rather than measured then the effectiveness of the earth must be confirmed by testing.

3. The value of prospective fault current recorded on the results schedule should be the greater of that measured between live conductors (short-circuit current) or phase conductor and earth (earth-fault current).

Figure 5.22 Schedule of Test Results

4. In the column headed 'Overcurrent Device', details of the main protective device including its short circuit should be inserted.

5. The column headed 'Reference Method' refers to the method used for the installation of cables also described in the *IEE On-Site Guide* (Table 7.1).

6. Continuity tests are to include a check for correct polarity of the following:

   - every fuse and single pole device to be connected in the phase conductor only

   - all centre contact lamp holders to have the outer contact connected to the neutral conductor

   - wiring of socket outlets and other accessories correctly connected.

7. Continuity of protective conductors – every protective conductor, including bonding conductors, must be tested to verify that it is sound and correctly connected.

8. Continuity of ring circuit conductors – a test must be carried out to verify the continuity of each conductor including the protective conductor of every ring circuit. A satisfactory reading should be indicated by a tick in the appropriate columns.

9. The sum of the resistances of the phase conductor and the protective conductor $(R_1 + R_2)$ should be recorded and, after appropriate correction for temperature rise, may be used to determine Z_s ($Z_s = Z_e + (R_1 + R_2)$). Where Test method 2 is used, the maximum value of R_2 should be recorded.

10. Insulation resistance – where necessary all electronic devices should be disconnected from the installation so that they are not damaged by the tests themselves. Details of such equipment should be recorded on the Schedule of Test Results so that the person carrying out routine periodic testing in the future is aware of the presence of such equipment.

11. The insulation resistance between conductors and between conductors and earth shall be measured and the minimum values recorded in the appropriate columns on the Schedule of Test Results. The minimum insulation resistance for the main switchboard and each distribution board tested separately with all final circuits connected but with all current-using equipment disconnected shall be not less than that given in the *IEE On-Site Guide* (Table 10.1).

12. Following the energising of the installation, polarity should be re-checked before further testing.

13. Earth-loop impedance (Z_s) – this may be determined either by direct measurement at the furthest point of a live circuit or by measuring Z_e at the origin of the circuit and adding to it the measured value of $R_1 + R_2$. Alternatively the value of Z_e may be obtained by inquiry from the electricity supply company. Values of Z_s should be less than the values (corrected where necessary) given in Appendix 2 of the *IEE On-Site Guide*.

14. Functional testing – the correct operation of all RCDs and RCBOs must be tested by using an instrument that simulates a fault condition. The time taken to trip the device is automatically recorded by the instrument and should be written in the appropriate space on the Schedule of Test Results. At rated tripping current RCDs should operate within 200 milliseconds.

Electrical Installation – Minor Works Certificate

Minor works are defined as an addition to an electrical installation that does not extend to the installation of a new circuit e.g. either the addition of a new socket outlet or of a lighting point to an existing circuit. This certificate includes space for the recording of essential test results and does not require the addition of a Schedule of Test Results.

Scope

The Minor Works Certificate shown in Figure 5.23 is only to be used for additions to an electrical installation that do not include the introduction of a new circuit e.g. the addition of an extra socket outlet or lighting point to an existing circuit (see BS 7671 Regulation 743-01-02).

Part 1 – Description

- Description of the minor works – this requires a full and accurate description in order that thc work carried out can be readily identified.

- Location/address – state the full address of the property and also where the work carried out can be located.

- Date of completion.

- Departures from BS 7671 (if any) – departures from the Regulations would only be expected in the most unusual circumstances.

Part 2 – Installation details

- Earthing arrangements – the type of supply should be indicated by a tick i.e. TN-C-S, TN-S, TT etc.

- Method of protection against indirect contact – the method of protection against indirect contact should be clearly identified e.g. earthed equipotential bonding and automatic disconnection of supply using fuse/circuit breaker or RCD.

- Protective device – state the type and rating of the protective device used.

- Comments on existing installation – if the existing installation lacks either an effective means of earthing or adequate main equipotential bonding conductors, this must be clearly stated. Any departures from BS 7671 may constitute non-compliance with either the *Electricity Supply Regulations* or the *Electricity at Work Regulations* and it is important that the client is informed in writing as soon as possible.

Figure 5.23   Minor Works Certificate

Part 3 – Essential tests

- Earth continuity – it is essential that the earth terminal of any new accessory is satisfactorily connected to the main earth terminal.
- Insulation resistance – the insulation resistance of the circuit that has been added to must comply with the requirements of BS 7671 (Table 71A).
- Earth-fault loop impedance – the earth-fault loop impedance must be measured to establish that the maximum permitted disconnection time is not exceeded.
- Polarity – a check must be made to ensure that the polarity of the additional accessory is correct.
- RCD operation – if the additional work is protected by an RCD, then it must be checked to ensure that it is working correctly.

Part 4 – Declaration

- The certificate must be completed and signed by a competent person in respect of the design, construction, and inspection and testing of the work carried out (in the case of minor works this will invariably be the responsibility of one person).
- The competent person must have a sound knowledge and experience relevant to the work undertaken and to the technical standards laid down in BS 7671. This person should also be fully aware of the inspection and testing procedures contained in the Regulations and must use suitable test instruments.
- When signing the form on behalf of a company or other organisation, individuals should state clearly whom they represent.

Guidance notes (for those receiving the certificate)

- This certificate has been issued to confirm that the work to which it relates has been designed, constructed, and inspected and tested in accordance with BS 7671.
- You should have received the original certificate and the contractor should have retained a duplicate. If you are not the owner of the installation then you should pass the certificate or a copy of it to the owner.
- The Minor Works Certificate is only to be used for additions, alterations or replacements to an installation that do not include the provision of a new circuit. This may include the installation of an additional socket outlet or lighting point to an existing circuit or the replacement or relocation of an accessory such as a light switch. A separate certificate should be issued for each existing circuit on which work of this nature has been carried out. The Minor Works Certificate is not valid where more extensive work is carried out, when an Electrical Installation Certificate should be issued.
- The original certificate should be retained in a safe place and be shown to any person inspecting the installation or carrying out further work in the future. If you later vacate the property, this certificate will demonstrate to the new owner that the minor works covered by the certificate met the requirements of the Regulations at the time the certificate was issued.

Initial inspection

The following text provides a detailed description of the procedures required to carry out an initial inspection of an electrical installation. Substantial reference has been made to the *IEE Wiring Regulations* (BS 7671), the *IEE On-Site Guide* and *IEE Guidance Note No.3* and it is recommended that wherever possible these documents are referred to should clarification be required.

The most important considerations prior to carrying out any inspection and test procedure are that all the required information is available, the person carrying out the procedure is competent to do so and that all safety requirements have been met. Forward planning is also a major consideration and it is essential that suitable inspection checklists have been prepared and that appropriate certification document is available for completion. It is also important to realise that a large proportion of any new installation will be hidden from view once the building fabric has been completed and therefore it is preferable to carry out a certain amount of visual inspection throughout the installation process, e.g. conduit, cable tray or trunking is often installed either above the ceiling or below the floor and once the ceiling or floor tiles have been fitted it is difficult and often expensive to gain access for inspection purposes.

The same applies to testing and it may be advisable to carry out tests such as earth continuity during construction rather than after the building has been completed.

It must be remembered, however, that when visual inspection and/or tests are carried out during the construction phase, the results must be recorded on the appropriate checklists or test certificates.

Information

Before carrying out the inspection and test of an installation it is essential that the person carrying out the work be provided with the following information:

1. The maximum demand of the installation expressed in amperes per phase together with details of the number and type of live conductors both for the source of energy and for each circuit to be used within the installation (e.g. single-phase two-wire a.c. or three-phase four-wire a.c. etc.).

2. The general characteristics of the supply such as:

   - the nominal voltage (U_o)
   - the nature of the current (I) and its frequency (Hz)
   - the prospective short circuit current at the origin of the installation (kA)
   - the earth-fault loop impedance (Z_e) of that part of the system external to the installation
   - the type and rating of the overcurrent device acting at the origin of the installation.

> **Remember**
>
> Although the major part of any inspection will be visual, other human senses may be employed e.g. a piece of equipment with moving parts may generate an unusual noise if it is not working correctly, or an electrical device which overheats will be hot to touch as well as giving off a distinctive smell. The senses of hearing, touch and smell will assist in detecting these

If this information is not known it must be established either by calculation, measurement, inquiry or inspection.

3. The type of earthing arrangement used for the installation e.g. TN-S, TN-C-S, TT etc.

4. The type and composition of each circuit (i.e. details of each sub-circuit, what it is feeding, the number and size of conductors and the type of wiring used).

5. The location and description of all devices installed for the purposes of protection, isolation and switching (e.g. fuses/circuit breakers etc.).

6. Details of the method selected to prevent danger from shock in the event of an earth fault. (This will invariably be protection by earthed equipotential bonding and automatic disconnection of the supply.)

7. The presence of any sensitive electronic device which may be susceptible to damage by the application of 500 volts d.c. when carrying out insulation resistance tests.

This information may be gained from a variety of sources such as the project specification, contract drawings, as fitted drawings or distribution board schedules. If such documents are not available, then the person ordering the work should be approached.

Scope

BS 7671 states that as far as reasonably practicable, an inspection shall be carried out to verify that:

- all equipment and materials used in the installation are of the correct type and comply with the appropriate British Standards or acceptable equivalent
- all parts of the installation have been correctly selected and installed
- no part of the installation is visibly damaged or otherwise defective
- the installation is suitable for the surrounding environmental conditions.

Inspection requirements

In order to meet the above requirements the inspection process should include the checking of at least the relevant items from the following list.

1. Connection of conductors

Every connection between conductors or between conductors and equipment must be electrically continuous and mechanically sound. We must also make sure that all connections are adequately enclosed but accessible where required by the Regulations.

2. Identification of conductors

A check should be made that each conductor is identified in accordance with the requirements of BS 7671 Table 51A and Table 51B. Although numbered sleeves or discs may be used in special circumstances, the most common form of identification is by means of coloured insulation or sleeving. It should be noted in particular that only protective conductors shall be identified by a combination of the colours green and yellow.

3. Cable routes

Cables should be routed out of harm's way and protected against mechanical damage where necessary. Permitted cable routes are clearly defined in the *On-Site Guide* or alternatively cables should be installed in earthed metal conduit or trunking.

4. Current carrying capacity

Where practicable the size of cable used should be checked for current carrying capacity and voltage drop based upon information provided by the installation designer.

5. Verification of polarity

A check must be made that all single pole devices are connected in the phase conductor only.

6. Accessories and equipment

Accessories and equipment should be checked to ensure they have been connected correctly including correct polarity.

7. Selection and erection to minimise the spread of fire

A check must be made (preferably during construction) that fire barriers, suitable seals and/or other means of protection against thermal effects have been provided as necessary to meet the requirements of the Regulations.

8. Protection against electric shock

A check must be made that the requirements of the Regulations have been met for the method of protection used. The most common of these are described below.

Protection against direct contact (defined as the contact of persons or livestock with live parts)

- Insulation – although protection by insulation is the usual method of protection against direct contact, other methods can be used. However, where insulation should be present it should be checked to ensure that no live conductors have been left exposed.

- Barriers – where live parts are protected by barriers or enclosures (e.g. bare busbars enclosed in a metal busbar chamber) these should be checked to ensure that all covers have been fitted and all fixing devices are secure.

- Obstacles – protection by obstacles provides protection only against unintentional contact with live conductors. If this method is used the area should be accessible only to skilled persons or persons under supervision.

- Out of reach – placing live parts out of reach can also provide protection against direct contact, although increased distances may be necessary where long or bulky conducting objects are likely to be handled in the vicinity. The requirements for this method are given more fully in Appendix 3 of the *Memorandum of Guidance to the Electricity at Work Act*.

Protection against indirect contact

Indirect contact is defined as the contact of persons or livestock with exposed conductive parts which have become live under fault conditions. Methods of protection against indirect contact are given in BS 7671 as:

- earthed equipotential bonding and automatic disconnection of supply
- use of class II equipment
- non-conducting location
- earth-free local equipotential bonding
- electrical separation.

Indirect contact is where persons or livestock come into contact with an exposed conductive part, which has become live under fault conditions. Examples of exposed conductive parts include metal trunking, metal conduit or exposed metal parts of an appliance such as an electric kettle.

Should the insulation of any of the live parts within the kettle become defective then the metal casing may become live and anyone touching the kettle would be at risk of receiving a dangerous electric shock. The most commonly used method of protection against indirect contact is earthing, equipotential bonding and automatic disconnection of the supply and it is these requirements that should be checked at the initial inspection stage.

Earthing arrangements: earthing conductors, main equipotential bonding conductors, circuit protective conductors and supplementary bonding conductors should all be checked to ensure that they have been correctly installed, are of the correct size and are correctly labelled.

Did you know?

Typical earthing arrangements for domestic installations are shown diagrammatically in the *On-Site Guide*; for other types of installation the appropriate size of earthing and bonding conductors should be determined in accordance with BS 7671 Chapter 54

Protection against both direct and indirect contact

Separate extra low voltage (SELV) is the most common method of providing protection against both direct and indirect contact. Requirements for this type of system include:

- an isolated source of supply e.g. a safety-isolating transformer to BS 3535 (also numbered BS EN 60742 1996)

- electrical separation, which means no electrical connection between the SELV circuit and higher voltage systems

- no connection with earth or the exposed conductive parts or protective conductors of other systems.

Specialised systems

Non-conducting locations and earth-free situations are specialised systems, which would only normally be used where specified and controlled by a suitably qualified electrical engineer. Although it is useful to be aware of the nature of these systems you are unlikely to be asked to carry out an inspection and test of these types of installation at this stage of your career.

9. Prevention of mutual detrimental influence

Account must be taken of the proximity of other electrical services of a different voltage band and of non-electrical services and influences, e.g. fire alarm and emergency lighting circuits must be separated from other cables and from each other, and Band 1 and Band 2 circuits must not be present in the same enclosure or wiring system unless they are either segregated or wired with cables suitable for the highest voltage present. Mixed categories of circuits may be contained in multicore cables, subject to certain requirements.

Band 1 circuits are circuits that are nominally extra-low voltage i.e. not exceeding 50 volts a.c. or 120 volts d.c., e.g. telecommunications or data and signalling. Band 2 circuits are circuits that are nominally low voltage, i.e. exceeding extra-low voltage but not exceeding 1000 volts a.c. between conductors or 600 volts a.c. between conductors and earth.

10. Isolating and switching devices

BS 7671 requires that effective means suitably positioned and ready to operate shall be provided so that all voltage may be cut off from every installation, every circuit within the installation and from all equipment, as may be necessary to prevent or remove danger. This means that switches and/or isolating devices of the correct rating must be installed as appropriate to meet the above requirements. It may be advisable where practicable to carry out an isolation exercise to check that effective isolation can be achieved. This should include switching off, locking-off and testing to verify that the circuit is dead and no other source of supply is present.

11. *Under voltage protection*

Sometimes referred to in starters as no-volt protection, suitable precautions must be taken where a loss or lowering of voltage or a subsequent restoration of voltage could cause danger. The most common situation would be where a motor driven machine stops due to a loss of voltage and unexpectedly restarts when the voltage is restored unless precautions such as the installation of a motor starter containing a contactor are employed. Regulations require that where unexpected restarting of a motor may cause danger, the provision of a motor starter designed to prevent automatic restarting must be provided.

12. *Labelling*

A check should be carried out to ensure that labels and warning notices as required by BS 7671 have been fitted e.g. labelling of circuits, MCBs, RCDs, fuses and isolating devices, periodic inspection notices advising of the recommended date of the next inspection, and warning notices referring to earthing and bonding connections.

13. *Selection of equipment appropriate to external influences*

All equipment must be selected as suitable for the environment in which it is likely to operate. Items to be considered are ambient temperature, presence of external heat sources, presence of water, likelihood of corrosion, ingress of foreign bodies, impact, vibration, flora, fauna, radiation, building use and structure.

14. *Access to switchgear and equipment*

BS 7671 requires that every piece of equipment that requires operation or attention must be installed so that adequate and safe means of access and working space are provided.

15. *Presence of diagrams, charts and other similar information*

All distribution boards should be provided with a distribution board schedule that provides information regarding types of circuits, number and size of conductors, type of wiring etc. These should be attached within or adjacent to each distribution board.

16. *Erection methods*

Correct methods of installation should be checked, in particular fixings of switchgear, cables, conduit etc., which must be adequate and suitable for the environment.

Inspection Schedule

All the previous items should be inspected and the results noted on an Inspection Schedule, an example of which is given in Figure 5.24. Where items are not applicable to the type of installation being inspected (e.g. different methods of protection against direct and indirect contact), these should either be ruled out or marked N/A (not applicable).

Inspection Schedule

1 Correction of conductors ☐

2 Identification of conductors ☐

3 Routing of cables in safe zones or protected against damage ☐

4 Selection of conductors for current and voltage drop ☐

5 Connection of single-pole devices in phase conductor only ☐

6 Correct connection of socket outlets and lamp holders ☐

7 Presence of fire barriers and protection against thermal effects ☐

8 Method of protection against shock ☐

8a Protection against both direct and indirect contact

SELV ☐

Limitation of discharge of energy ☐

8b Protection against direct contact

Insulation of live parts ☐

Batteries or enclosures ☐

Obstacles ☐

Placing out of reach ☐

PELV ☐

8c Protection against indirect contact ☐

8d (i) Earthed equipotential bonding and automatic disconnection ☐

Presence of bonding conductors ☐

Presence of earthing conductors ☐

Presence of protective conductors ☐

8d (ii) Use of class 2 equipment ☐

8d (iii) Non-conducting locations ☐

8d (iv) Earth-free local equipotential bonding ☐

Presence of earth-free equipotential bonding conductors ☐

8d (v) Electrical Separation ☐

9 Prevention of mutual detrimental influence

Priority of non-electrical services ☐

Separation of Band 1 and Band 2 circuits ☐

Separation of fire alarm and emergency lighting circuits ☐

10 Presence of appropriate devices for isolation and switching ☐

11 Presence of under voltage protection devices ☐

12 Choice and setting of protective and monitoring devices
(a) residual current devices
(b) overcurrent device ☐

13 Labelling of protective devices switches and terminals ☐

14 Selection of equipment suitable for external devices ☐

15 Adequate access to switchgear and equipment ☐

16 Presence of danger notices and other warning signs ☐

17 Presence of diagrams and instructions ☐

18 Erection methods ☐

19 Requirements for special locations ☐

Signed: _____

Date: _____

Figure 5.24 Inspection Schedule

Inspection checklists

To ensure that all the requirements of the Regulations have been met, inspection checklists should be drawn up and used appropriate to the type of installation being inspected. Examples of suitable checklists are given in Table 5.9 which follows.

Joint Boxes (tick if satisfactory)	
All joint boxes comply with the appropriate British Standard.	
Joints accessible for inspection where required.	
All conductors correctly connected.	
Joints protected against mechanical damage.	

Wiring Accessories (General Requirements) (tick if satisfactory)	
All accessories comply with the appropriate British Standard.	
Boxes and other enclosures securely fastened.	
Metal boxes and enclosures correctly earthed.	
Flush boxes not projecting above surface of wall.	
No sharp edges which could cause damage to cable insulation.	
Non-sheathed cables not exposed outside box or enclosure.	
Conductors correctly identified.	
Bare protective conductors sleeved green and yellow.	
All terminals tight and contain all strands of stranded conductor.	
Cord grips correctly used to prevent strain on terminals.	
All accessories of adequate current rating.	
Accessories suitable for all conditions likely to be encountered.	

Socket Outlet (tick if satisfactory)	
Comply with appropriate British Standard.	
Mounting height above floor or working surface is correct.	
All sockets have correct polarity.	
Sockets not installed in bathroom or shower room (unless shaver type socket).	
Sockets not within 2.5 m of a shower in a room other than a bathroom.	
Sockets controlled by a switch if the supply is direct current.	
Sockets correctly protected where floor mounted.	
Circuit protective conductor connected directly to the earthing terminal of the socket outlet on a sheathed wiring installation.	

Table 5.9 Inspection checklists (continued overleaf)

Socket Outlet continued (tick if satisfactory)

Earthing tail provided from the earthed metal box to the earthing terminal of the socket outlet.	
Socket outlets not used to supply a water heater with un-insulated elements.	

Lighting Controls (tick if satisfactory)

Light switches comply with appropriate British Standard.	
Switches suitably located.	
Single-pole switches connected in phase conductor only.	
Correct colour coding of conductors.	
Correct earthing of metal switch plates.	
Switches out of reach of a person using bath or shower.	
Switches for inductive circuits (discharge lamps) de-rated as necessary.	
Switches labelled to indicate purpose where this is not obvious.	
All switches of adequate current rating.	
All controls suitable for their associated luminaire.	

Lighting Points (tick if satisfactory)

All lighting points correctly terminated in suitable accessory or fitting.	
Ceiling roses comply with appropriate British Standard.	
No more than one flexible cord unless designed for multiple pendants.	
Devices provided for supporting flex used correctly.	
All switch wires identified.	
Holes in ceiling above ceiling rose made good to prevent spread of fire.	
Ceiling roses not connected to supply exceeding 250 V.	
Flexible cords suitable for the mass suspended.	
Lamp holders comply with appropriate British Standard.	
Luminaire couplers comply with appropriate British Standard.	

Conduits (general) (tick if satisfactory)

All inspection fittings accessible.	
Maximum number of cables not exceeded.	
Solid elbows used only as permitted.	
Conduit ends reamed and bushed.	
Adequate number of boxes.	

Table 5.9 Inspection checklists (continued overleaf)

Conduits (general) continued (tick if satisfactory)	
All unused entries blanked off.	
Lowest point provided with drainage holes where required.	
Correct radius of bends to prevent damage to cables.	
Joints and scratches in metal conduit protected by painting.	

Circuit Protective Conductors (enter circuit details from specifications)	
1.	
2.	
3.	
4.	
5.	
6.	
7.	
8.	

Table 5.9   Inspection checklists (continued)

Did you know?

Where an item is not relevant to a specific installation N/A should be placed in the tick box

Periodic inspection and testing

General requirements

The regular inspection and testing of electrical installations is necessary because over a period of time the condition of all installations will deteriorate to some extent. This may be due to normal wear and tear, accidental damage, corrosion or other effects due to environmental influences, normal ageing or deterioration due to excessive electrical loading.

The Electricity at Work Regulations (1989) state that: 'As may be necessary to prevent danger, all systems shall be maintained so as to prevent, as far as is reasonably practicable, such danger'.

This means that all electrical installations must be maintained in a safe condition, and regular inspection and testing (periodic inspection) is an essential part of any such preventative maintenance programme. In addition to statutory requirements other bodies such as licensing authorities, insurance companies, mortgage lenders etc. may also require periodic inspection and testing to be carried out on a regular basis.

Other reasons for carrying out periodic inspection and testing are:

- to confirm compliance with BS 7671
- on a change of ownership of the premises
- on a change of use of the premises
- on a change of tenancy of the premises
- on completion of alterations or additions to the original installation
- following any significant increase in the electrical loading of the installation
- where there is reason to believe that damage may have been caused to the installation.

In the case of an installation that is under constant supervision while in normal use, such as a factory or other industrial premises, periodic inspection and testing may be replaced by a system of continuous monitoring and maintenance of the installation provided that adequate records of such maintenance are kept.

When carrying out the design of an electrical installation, and particularly when specifying the type of equipment to be installed, the designer should take into account the likely quality of the maintenance programme, and the periods between periodic inspection and testing to be specified on the Electrical Installation Certificate.

Both Section 6 of the Health and Safety at Work Act and the *Construction (Design and Management) Regulations* require information on the requirements for routine checks and periodic inspections to be provided. The advice of the Health and Safety

Did you know?

The requirements for periodic inspection and testing are given in BS 7671 Chapter 73

Executive in their *Memorandum of Guidance on the Electricity at Work Regulations* indicates that practical experience of an installation's use may indicate a need for an adjustment to the frequency of checks and inspections i.e. more often or less frequent depending on the likely deterioration of the installation during its normal use. This would be a matter of judgement for the duty holder.

Routine checks

Electrical installations should still be routinely checked in the intervening time between periodic inspection and testing. In domestic premises it is likely that the occupier will soon notice any damage or breakages to electrical equipment and will take steps to have repairs carried out. In commercial or industrial installations a suitable reporting system should be available for users of the installation to report any potential danger from deteriorating or damaged equipment.

In addition to this, a system of routine checks should be set up to take place between formal periodic inspections. The frequency of these checks will depend entirely on the nature of the premises and the usage of the installation. Routine checks are likely to include activities such as those listed in Table 5.10.

Activity	Check
Defect reports	Check that all reported defects have been rectified and that the installation is safe
Inspection	Look for: Breakages Wear or deterioration Signs of overheating Missing parts (covers/screws) Switchgear still accessible Enclosure doors secure Labels still adequate (readable) Loose fittings
Operation	Check operation of: Switchgear (where reasonable) Equipment (switch off and on) RCD (using test button)

Table 5.10 Routine checks

The recommended period between both routine checks and formal inspections are given in Table 3.2 taken from *IEE Guidance Note 3*. The requirements for such inspections are stated in BS 7671 Chapter 73 and specify that all inspections should provide careful scrutiny of the installation without dismantling or with only partial dismantling where absolutely necessary. It is considered that the unnecessary dismantling of equipment or disconnection of cables could produce a risk of introducing faults that were not there in the first place.

The frequency of periodic inspection and testing should aim to provide, as far as reasonably possible, the following:

- the safety of persons and livestock against the effects of electric shock or burns
- protection against damage to property by fire or heat arising from an installation defect
- confirmation that the installation has not been damaged and has not deteriorated to the extent that it may impair safety
- the identification of any defects in the installation or non-conformity with the current edition of the Regulations that may cause danger.

In summary, the inspection should ensure that:

- the installation is safe
- the installation has not been damaged
- the installation has not deteriorated so as to impair safety
- any items that no longer comply with the Regulations or may cause danger are identified.

In practical terms the inspector is carrying out a general inspection to ensure that the installation is safe. However, the inspector is required to record and make recommendations with respect to any items that no longer comply with the current edition of the Regulations.

As with all inspections the person carrying out the work must be competent and have sufficient knowledge and experience of the type of installation to be inspected and tested. Enquiries should be made to the person responsible for the installation with regard to the provision of charts and/or diagrams to indicate the type of circuits, means of isolation and switching, and types and ratings of protective devices including a written record of all previous inspection and test results.

Statutory and non-statutory documentation

1. *The Electricity Supply Regulations (1988).*

2. *The Electricity at Work Regulations.*

3. BS 5266 Pt.1 Code of Practice for emergency lighting systems (other than cinemas). Other Regulations and intervals cover testing of batteries and generators.

4. BS 5839 Pt.1 Code of Practice for the design, installation and servicing of fire alarm systems.

5. Local authority conditions of licence.

6. SI 1995 No 1129 (clause 27) *The Cinematography (Safety) Regulations.*

Guidance Note 3 requirements for inspection and testing

Guidance Note 3 3.8.1 General Procedure

In old installations where information such as drawings, distribution board schedules, charts etc. are not available, some exploratory work may be necessary to enable inspection and testing to be carried out safely and without damage to existing equipment. A survey should be carried out to identify all items of switchgear and control gear and their associated circuits.

During the survey a note should be made of any known changes in either the structure of the building, the environmental conditions or of any alterations or additions to the installation which may affect the suitability of the wiring or the method of installation.

A careful check should be made of the type of equipment on site so that, where required, electronic or other equipment that may be damaged by high-test voltages can be either disconnected or short-circuited to prevent damage. If computer equipment is to be disconnected for the purpose of testing then two important issues need to be considered:

- The user of the equipment must be informed in order that data within the computer can be backed up and stored on disc if necessary.

- Many commercial computer installations have an emergency back-up electrical supply that will automatically energise if the mains supply is disconnected. This means that circuits that you assume are isolated may well be kept live from a different source of supply.

If the inspection and testing cannot be carried out safely without the use of drawings, diagrams or schedules etc. then the person ordering the work should be informed.

Guidance Note 3 3.8.2

The requirements of BS 7671 for Periodic Inspection and Testing are for a thorough **Inspection** of the installation supplemented by **Testing** where necessary. For safety reasons a visual inspection must be carried out before opening any enclosures, removing covers or carrying out any tests.

To comply with the Electricity at Work Regulations (1989) the inspection should be carried out wherever possible with the supply de-energised. A thorough visual inspection should be carried out of all electrical equipment that isn't concealed and, where damage or deterioration has occurred, this must be recorded on the inspection schedule. The inspection should include a check on the condition of all equipment and materials used in the installation with regard to the following:

- safety
- damage
- external influences
- wear and tear
- overload
- suitability
- corrosion
- age
- correct operation.

Detailed inspection requirements

Anticipation of danger

As stated previously, when carrying out the inspection the opportunity should be taken to identify any equipment that may be damaged if subjected to high-test voltages. As well as computer equipment this may include safety systems such as fire or intruder alarms that could well have electronic components susceptible to test voltages.

Joints and connections

It may be impossible to inspect every joint and termination in an electrical installation. Therefore, where necessary, a sample inspection should be made. Provided the switchgear and distribution boards are accessible as required by the Regulations, then a full inspection of all conductor terminations should be carried out and any signs of overheating or loose connections should be investigated and included in the report. For lighting points and socket outlets a suitable sample should be inspected in the same way.

Conductors

The means of identification of every conductor including protective conductors should be checked and any damage or deterioration to the conductors, their insulation, protective sheathing or armour should be recorded. This inspection should include each conductor at every distribution board within the installation and a suitable sample of lighting points, switching points and socket outlets.

Flexible cables and cords

Where flexible cables or cords form part of the fixed installation the inspection should include:

- examination of the cable or cord for damage or deterioration
- examination of the terminations and anchor points for any defects
- checking the correctness of the installation with regard to additional mechanical protection or the application of heat resistant sleeving where necessary.

Switches

The *IEE Guidance Notes 3 (Inspection & Testing)*, recommends that a random sample of at least 10 per cent of all switching devices be given a thorough internal visual inspection to assess their electrical and mechanical condition. Should the inspection reveal excessive wear and tear or signs of damage due to arcing or overheating then, unless it is obvious that the problem is associated with that particular switch, the inspection should be extended to include all remaining switches associated with the installation.

Protection against thermal effects

Although sometimes difficult due to the structure of the building, the presence of fire barriers and seals should be checked wherever reasonably practicable.

Protection against direct and indirect contact

Separate Extra Low Voltage (SELV) is commonly used as a means of protection against both direct and indirect contact. When inspecting this type of system, the points to be checked include the use of a safety isolating transformer, the need to keep the primary and secondary circuits separate and the segregation of exposed conductive parts of the SELV system from any connection with the earthing of the primary circuit or from any other connection with earth.

Protection against direct contact

Inspection of the installation should confirm that all the requirements of the Regulations have been met with regard to protection against direct contact with live conductors. This means checking to ensure there has been no damage or deterioration of any of the insulation within the installation, no removal of barriers or obstacles and no alterations to enclosures that may allow access to live conductors.

Protection against indirect contact

The method used for protection against indirect contact must be established and recorded on the Inspection Schedule. Where earthed equipotential bonding and automatic disconnection of the supply is used, a check on the condition of the main equipotential bonding conductor and the satisfactory connection of all other protective conductors with earth are essential.

Protective devices

A check must be made that each circuit is adequately protected with the correct type, size and rating of fuse or circuit breaker. A check should also be made that each protective device is suitable for the type of circuit it is protecting and the earthing system employed, e.g. will the protective device operate within the disconnection time allowed by the Regulations and is the rating of the protective device suitable for the maximum prospective short circuit current likely to flow under fault conditions?

Enclosures and mechanical protection

The enclosures of all electrical equipment and accessories should be inspected to ensure that they provide protection not less than IP2X or IPXXB, and where horizontal top surfaces are readily accessible they should have a degree of protection of at least IP4X. IP2X represents the average finger of 12.5 mm diameter and 80 mm in length and can be tested by a metal finger of these dimensions. IP4X provides protection against entry by strips greater than 1.0 mm or solid objects exceeding 1.0 mm in diameter.

> **Remember**
>
> An RCD must not be used as the sole means of protection against direct contact with live parts

Marking and labelling

Labels should be applied adjacent to every fuse or circuit breaker indicating the size and type of fuse or the nominal current rating of the circuit breaker and details of the circuit they protect. Other notices and labels required by the Regulations are as follows:

1. At the origin of every installation as shown in Figure 5.25.

> **IMPORTANT**
> This installation should be periodically inspected and tested and a report on its condition obtained as prescribed in BS7671 (formally the IEE Wiring Regulations for Electrical Installations) published by the Institute of Electrical Engineers.
>
> Date of last inspection : ....
>
> Recommended date of next inspection : ....

Figure 5.25   Label for origin of installation

2. Where different voltages are present:

   - in equipment or enclosures within which a voltage exceeding 250 volts exists but where such a voltage would not be expected

   - terminals between which a voltage excceding 250 volts exists, which although contained in separate enclosures are within arm's reach of the same person

   - the means of access to all live parts of switchgear or other live parts where different nominal voltages exist.

3. Earthing and bonding connections: a label with the words shown in Figure 5.26 shall be permanently fixed in a visible position at or near the following points:

   - the point of connection of every earthing conductor to an earth electrode

   - the point of connection of every bonding conductor to an extraneous conductive part

   - the main earth terminal of the installation where separate from the main switchgear.

> **Safety Electrical Connection
> – Do Not Remove**

Figure 5.26   Label for earthing and bonding connections

4. Residual Current Devices: where RCDs are fitted within an installation, a suitable permanent durable notice marked in legible type no smaller than the example shown in BS 7671 shall be permanently fixed in a prominent position at or near the main distribution board. The notice shall contain the following words:

   'This installation, or part of it, is protected by a device that automatically switches off the supply if an earth-fault develops. Test quarterly by pressing the button marked 'T' or 'Test'. The device should switch off the supply and should then be switched on to restore the supply. If the device does not switch off the supply when the button is pressed, seek expert advice.'

Caravan installations: all touring (mobile) caravans shall have a notice fixed near the main switch giving instructions on the connection and disconnection of the caravan installation to the electricity supply. Details of the wording required on the notice are given in full in BS 7671.

Periodic testing

Periodic testing is supplementary to periodic inspection and the same level of testing as for a new installation is not necessarily required.

When the building being tested is unoccupied, isolation of the supply for testing purposes will not cause a problem, but where the building is occupied, then testing should be carried out to cause as little inconvenience as possible to the user. Where isolation of the supply is necessary then the disconnection should take place for as short a time as possible and by arrangement with the occupier. It may be necessary to complete the tests out of normal working hours in order to keep disruption to a minimum.

Sequence of tests

Because it may not be possible to disconnect the installation under test from the supply, the recommended sequence of carrying out periodic tests is different from the sequence of tests for a new installation and is given below:

- continuity of all protective conductors (including equipotential bonding conductors and continuity of ring circuit conductors where required)
- polarity
- earth-fault loop impedance
- insulation resistance
- operation of devices for isolation and switching
- operation of residual current devices
- operation of circuit breakers.

Detailed test procedures

Continuity of protective conductors and equipotential bonding conductors

Where the installation can be safely isolated from the supply, then the circuit protective conductors and equipotential bonding conductors can be disconnected from the main earthing terminal in order to verify their continuity.

Where the installation cannot be isolated from the supply, the circuit protective conductors and the equipotential bonding conductors must not be disconnected from the main earthing terminal as under fault conditions extraneous metalwork could become live. Under these circumstances a combination of inspection, continuity testing and earth loop impedance testing should establish the integrity of the circuit protective conductors.

When testing the effectiveness of the main bonding conductors or supplementary bonds, the resistance value between any service pipe or extraneous metalwork and the main earthing terminal should not exceed 0.05 ohms.

Polarity

Polarity tests should be carried out to check that:

- polarity is correct at the intake position and the consumer unit or distribution board. Single pole switches or control devices are connected in the phase conductor only
- socket outlets and other accessories are connected correctly
- centre contact bayonet and Edison screw type lamp holders have their outer or screwed contact connected to the neutral conductor
- all multi-pole devices are correctly installed.

Where it is known that no alterations or additions have been made to the installation since its last inspection and test, then the number of items to be tested can be reduced by sampling. It is recommended that at least 10 per cent of all switches and control devices should be tested together with any centre contact lamp holders and 100 per cent of socket outlets. However if any cases of incorrect polarity are found then a full test should be made of all accessories in that part of the installation and the sample of the remaining should be increased to 25 per cent.

Earth-fault loop impedance

Earth-fault loop impedance tests should be carried out at:

- the origin of each installation and at each distribution board
- all socket outlets
- at the furthest point of each radial circuit.

Results obtained should be compared with the values documented during previous tests, and where an increase in values has occurred these must be investigated.

Insulation resistance

Insulation resistance tests can only be carried out where it is possible to safely isolate the supply. All electronic equipment susceptible to damage should be disconnected or alternatively the insulation resistance test should be made between phase and neutral conductors connected together and earth. Where practicable the tests should be carried out on the whole of the installation with all switches closed and all fuse links in place. Where this is not possible, the installation may be subdivided by testing each separate distribution board one at a time.

BS 7671 Table 71A states a minimum acceptable resistance value of 0.5 megohms. However, if the measured value is less than 2 megohms then further investigation is required to determine the cause of the low reading.

Where individual items of equipment are disconnected for these tests and the equipment has exposed conductive parts, then the insulation resistance of each item of equipment should be checked. In the absence of any other requirements the minimum value of insulation resistance between live components and the exposed metal frame of the equipment should be not less than 0.5 megohms.

Operation of main switches and isolators

All main switches and isolators should be inspected for correct operation and clear labelling and to check that access to them has not been obstructed. Where the operation of the contacts of such devices is not visible it may be necessary to connect a test lamp between each phase and the neutral on the load side of the device to ensure that all supply conductors have been broken.

Operation of Residual Current Devices

RCDs should be tested for correct operation by the use of an RCD tester to ensure they trip out in the time required by BS 7671 as well as by use of the integral test button. A check should also be made that the tripping current for the protection of a socket outlet to be used for equipment outdoors should not exceed 30 mA.

Operation of circuit breakers

All circuit breakers should be inspected for visible signs of damage or damage caused by overheating. Where isolation of the supply to individual sub-circuits will not cause inconvenience, each circuit breaker should be manually operated to ensure that the device opens and closes correctly.

Periodic Inspection Report

BS 7671 requires that the results of any periodic inspection and test should be recorded on a Periodic Inspection and Test Report of the type illustrated in Figure 5.27. The report should include the following:

- a description of the extent of the inspection and tests and what parts of the installation were covered

- any limitations (e.g. portable appliances not covered)
- details of any damage, deterioration or dangerous conditions which were found
- any non-compliance with BS 7671
- schedule of test results.

Details of the Client

Name : . . . .
Address : . . . .
. . . .
. . . .

Purpose of Report : . . . .

Details of the Installation
Occupier : . . . .
Installation address (if different from above) : . . . .
. . . .
Description of Premises :   Domestic ☐   Commercial ☐   Industrial ☐   Other ☐
Estimated age of the installation : . . . . Years.   Evidence of alterations or additions : . . . . Yes/No
Date of last inspection : . . . .   Records available : . . . . Yes/No

Extent & Limitations of the Inspection
Extent of the installation covered by this report : . . . .
. . . .
Limitations : . . . .
. . . .
. . . .
This inspection has been carried out in accordance with BS7671 (I.E.E Wiring Regulations). Cables concealed within trunking and conduit or cables concealed under floors, buried underground, installed in roof spaces or generally hidden within the fabric of the building have not been inspected.

Next Inspection
I/We recommend that this installation should be further inspected and tested after an interval of not more than. . . . . Months/years, providing that any observations requiring urgent attention are attended to without delay.

Declaration
Inspected & tested by

Name : . . . .   Signature : . . . .
For and on behalf of : . . . .   Position : . . . .
Address : . . . .
. . . .
. . . .   Date : . . . .

Figure 5.27   Periodic Inspection Report

If any items are found which may cause immediate danger these should be rectified immediately. If this is not possible then they must be reported to a responsible person without delay.

When inspecting older installations, which may have been installed in accordance with a previous edition of the *IEE Wiring Regulations*, provided that all items which do not conform to the present edition of BS 7671 are reported, the installation may still be acceptable provided that no risk of shock, fire or burns exists.

Supply characteristics

Earthing Arrangements	Number and Type of Live Conductors	Nature of Supply	Protective Device Details
TN-C ☐	a.c 1-phase-2 wire ☐	Nominal voltage : V	Type : . . . .
TN-S ☐	1-phase-3 wire ☐	Nominal frequency :	. . . .
TN-C-S ☐	2-phase-3 wire ☐	. . . . .Hz	. . . .
TT ☐	3-phase-3 wire ☐	Prospective fault	Nominal current rating
Alternative (details to	3-phase-4 wire ☐	Current : . . . . .kA	. . . .
be given on attached	d.c 2-pole ☐	External loop	. . . . .A
schedules).	3-pole ☐	Impedance : . . . .ohms	

Particulars of the Installation

Means of Earthing   Details of the installation earth electrode

Suppliers facility ☐   Type (rod – tape etc)   Location   Resistance to

earth ☐

Earth electrode . . . . .ohms

Main Protective Conductors

Main earthing conductor : material . . . . .csa

Main equipotential bonding conductor : material . . . . .csa

To incoming water service . . . . ☐   To incoming gas service . . . . ☐

To incoming oil service . . . . ☐   To structural steel . . . . ☐

To lightning protection . . . . ☐

To other incoming services (state details) . . . .

Main Switch or Circuit Breaker

BS Type ; . . . .   No of poles : . . . .

Current rating : . . . . .A   Voltage rating : V

Location : . . . .   Fuse rating or setting : . . . . .A

Residual current device – operating current : . . .mA   Operating time : . . . . .ms

Figure 5.28 Supply characteristics and earthing arrangements

Summary of periodic testing

Type of test	Recommendations
Continuity	Tests to be carried out between: • all main bonding connections • all supplementary bonding connections Note! When an electrical installation cannot be isolated, protective conductors including bonding conductors must not be disconnected
Polarity	Tests to be carried out: • origin of installation • all socket outlets • 10 per cent of control devices (including switches) • 10 per cent of centre contact lamp holders Note! If incorrect polarity is found then a full test should be made on that part of the installation and testing on the remainder increased to 25 per cent. If further faults are found the complete installation must be tested
Earth-fault loop impedance	Tests to be carried out at: • origin of installation • each distribution board • each socket outlet • extremity of every radial circuit
Insulation resistance	If this test is to be carried out then test: • the whole installation with all protective devices in place and all switches closed • where electronic devices are present, the test should be carried out between phase and neutral conductors connected together and earth
Functional	Activities to be carried out: • all isolation and switching devices to be operated • all labels to be checked • all interlocking mechanisms to be verified • all RCDs to be checked both by test instrument and by test button • all manually operated circuit breakers to be operated to verify they open and close correctly

Table 5.11 Summary of periodic testing

On the job: Periodic inspection

You are asked to carry out a periodic inspection at a small workshop. All goes well until you come to check the office that is attached to the factory, where you notice that about 20 staff are working on computers. No installation drawings exist for the office, as it turns out the factory was built as an extension to the offices many years ago.

1   Can you identify the problems you will encounter in this job?

2   How would you deal with the situation?

(A suggested answer can be found on page 315)

Unsatisfactory test results

Continuity

When testing the continuity of circuit protective conductors or bonding conductors we should always expect a very low reading, which is why we must always use a low-reading ohmmeter. Main and supplementary bonding conductors should have a reading of not more than 0.05 ohms whilst the maximum resistance of circuit protective conductors can be estimated from the value of $(R_1 + R_2)$ given in Table 9A of the *IEE On-Site Guide*. These values will depend upon the cross-sectional area of the conductor, the conductor material and its length.

A very high (end of scale) reading would indicate a break in the conductor itself or a disconnected termination that must be investigated. A mid-range reading may be caused by the poor termination of an earthing clamp to the service pipe e.g. a service pipe which is not cleaned correctly before fitting the clamp or corrosion of the metal service pipe due to its age and damp conditions.

Polarity

Correct polarity is achieved by the correct termination of conductors to the terminals of all equipment. This may be main intake equipment such as isolators, main switches and distribution boards or accessories such as socket outlets, switches or lighting fittings. Polarity is either correct or incorrect; there is nothing in between. Incorrect polarity is caused by the termination of live conductors to the wrong terminals and is corrected by re-connecting all conductors correctly.

Insulation resistance

The value of insulation resistance of an installation will depend upon the size and complexity of the installation and the number of circuits connected to it. When testing a small domestic installation you may expect an insulation resistance reading in excess of 200 megohms whilst a large industrial or commercial installation with many sub-circuits, each providing a parallel path, will give a much smaller reading if tested as a whole. It is recommended that, where the insulation resistance reading is less than 2 megohms, individual distribution boards or even individual sub-circuits be tested separately in order to identify any possible cause of poor insulation values.

An extremely low value of insulation resistance would indicate a possible short circuit between live conductors or a bare conductor in contact with earth at some point in the installation, either of which must be investigated. A reading below 0.5 megohms would suggest a weakness in the insulation, possibly due to the ingress of dampness or dirt in such items as distribution boards, joint boxes or lighting fittings etc. Although PVC insulated cables are not generally subject to a deterioration of insulation resistance due to dampness (unless the insulation or sheath is damaged), mineral insulated cables can be affected if dampness has entered the end of a cable before the seal has been applied properly. Other causes of low insulation resistance can be the infestation of equipment by rats, mice or insects.

Did you know?

A reading only slightly higher than the required reading may be possible to correct by replacing the conductor with one of a larger cross-sectional area

Earth-fault loop impedance

As explained previously the earth-fault loop path is made up of those parts of the supply system external to the premises being tested (Z_e) and the phase conductor and circuit protective conductor within the installation ($R_1 + R_2$), the total earth-fault loop impedance being $Z_s = Z_e + (R_1 + R_2)$.

Should the value of impedance measured be higher than that required by the design of the installation, then as we have no influence on the external value of impedance (Z_e) we can only reduce the value of Z_s by installing circuit protective conductors of a larger cross-sectional area or, if aluminium conductors have been used, by changing these to copper. If the value were still too high to guarantee operation of the circuit protective device in the time required by BS 7671, then consideration would have to be given to changing the type of protective device (i.e. fuses to circuit breakers).

Residual Current Devices (RCDs)

Where a Residual Current Device (RCD) fails to trip out when pressing the integral test button this would indicate a fault within the device itself, which should therefore be replaced.

Where a Residual Current Device fails to trip out when being tested by an RCD tester then it would suggest a break in the earth return path, which must be investigated. If the RCD does trip out but not within the time specified then a check should be made that the test instrument is set correctly for the nominal tripping current of the device under test.

Remember

Further information and advice on the interpretation of test results and fault finding is given in Chapter 6 Fault diagnosis and rectification

FAQ

Q   When should I prove that my test instrument is working correctly?

A   You can never tell when something will stop working properly. Therefore, you should always prove that your test instrument is working correctly both before you start to test and after you have completed the test. Only then can you be sure that your readings are valid.

Q   Why do I need to do a ring circuit continuity test?

A   The purpose of this test is to establish that a ring exists and that it has been correctly connected. As the IEE Regulations require that each leg of the ring be tested, it is probably sensible to mark each conductor at the distribution board as being either an Incoming leg from the last socket, or an Outgoing leg going to the first.

FAQ

Q When carrying out a ring circuit continuity test, why is it that a higher reading at a socket indicates a spur exists?

A We know when we do this test, that for a correctly connected circuit, the reading at each socket should be roughly the same. Try to imagine a ring circuit for what it actually is, i.e. each conductor of the ring circuit leaving the distribution board terminal, travelling out around the various socket outlets, thus forming a big circle (ring) and then returning to that same distribution board terminal.

If you think about it, it doesn't matter where you are on a circle, the distance all the way around it back to where you are is the same, no matter where you start measuring it from. This is why the reading at each socket outlet should be roughly the same. If you are moving from socket to socket, obtaining similar readings and then reach a socket where the reading obtained is higher, you have probably found a spur. This is because the spur is connected to that socket by using additional cable outside the ring and you are now measuring the additional resistance of that cable, hence the higher reading.

Knowledge check

1. How often is it recommended that a hospital has an electrical inspection?

2. There are four reasons why we carry out an initial inspection. What are they?

3. Name five things that should be covered by an initial inspection.

4. Name all the initial tests that should be carried out on an electrical installation, placing them in the correct sequence and indicating which are the 'live' tests.

5. Describe what is meant by the term 'indirect contact'.

6. When carrying out an insulation resistance test, why is it important to remove all GLS lamps before applying the test?

7. Why is it unacceptable to only test an RCD by pressing the test button located on the front of the RCD?

8. A 30 mA RCD is providing supplementary protection against direct contact. A test current of what should cause the RCD to trip within how many milliseconds?

9. What is a Minor Works Certificate and when should it be used?

10. Describe a TN-C-S earthing system.

Chapter 6

Fault diagnosis and rectification

OVERVIEW

Electricians should have the ability to recognise when something is not up to standard or is not functioning correctly, such as when a piece of equipment is not suitable (e.g. a metal-clad switch that has been installed in a damp environment) or a circuit is not operating correctly (e.g. a two-way lighting circuit that is only operating from one switch). Not all faults are easily visible; some are concealed and some may develop over a period of time. Regular inspection, tests and maintenance checks will limit such faults. Testing is not only needed at the completion of works but should also take place during the installation process of the wiring system. This chapter will cover the following areas:

- **Safe working practices for fault diagnosis**

- **General categories of fault**

- **Specific types of fault**

- **Knowledge and understanding of electrical systems and equipment**

- **Stages of diagnosis and rectification**

- **Factors affecting the fault repair process**

- **Special situations**

Safe working practices for fault diagnosis

Isolation

Remember, that before beginning work on any piece of electrical equipment or circuit you should make sure that it is completely isolated from the supply by following recognised procedures.

One such procedure has been drawn up by the Joint Industry Board for the Electrical Contracting Industry and sets the standard for safe systems of working in the industry. The JIB procedure has two parts to it as shown in Figures 6.1 and 6.2:

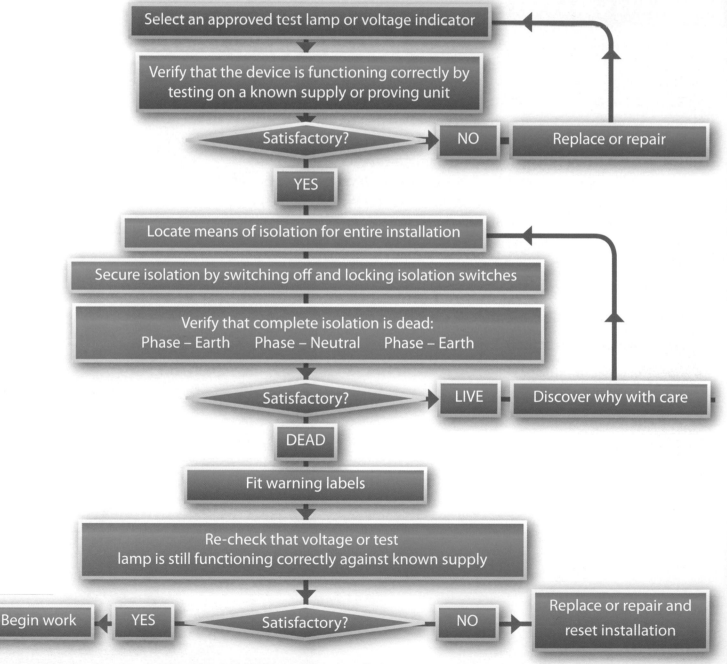

Figure 6.1   Isolating a complete installation

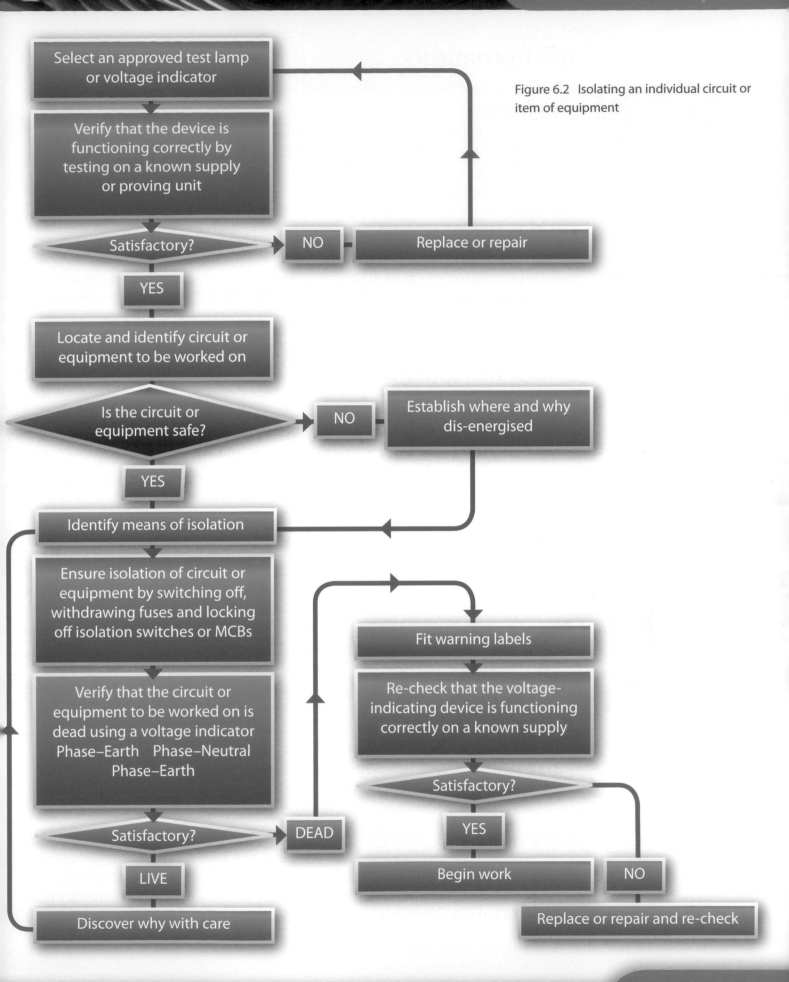

Figure 6.2 Isolating an individual circuit or item of equipment

Test instruments

All test equipment must be regularly checked to make sure it is in good and safe working order.

If you are checking the equipment yourself, the following points should be noted and, if you have any doubt about an instrument or its accuracy, ask for assistance, as test instruments are very expensive and any unnecessary damage caused by ignorance should be avoided.

Guidance Note GS 38

Published by the Health and Safety Executive, GS 38 is for electrical test equipment used by electricians and gives guidance to electrically competent people involved in electrical testing, diagnosis and repair. Electrically competent people may include electricians, electrical contractors, test supervisors, technicians, managers or appliance repairers.

Voltage indicating devices

Instruments used solely for detecting a voltage fall into two categories:

- Detectors that rely on an illuminated lamp (test lamp) or a meter scale (test meter): test lamps are fitted with a 15 watt lamp and should not give rise to danger if the lamp is broken. A guard should also protect it.

- Detectors that use two or more independent indicating systems (one of which may be audible) and limit energy input to the detector by the circuitry used. An example is a two-pole voltage detector, i.e. a detector unit with an integral test probe, an interconnecting lead and a second test probe.

Both these detectors are designed and constructed to limit the current and energy that can flow into the detector. This limitation is usually provided by a combination of the circuit design using the concept of protective impedance, and current-limiting resistors built into the test probes. The detectors are also provided with in-built test features to check the functioning of the detector before and after use. The interconnecting lead and second test probes are not detachable components. These types of detector do not require additional current-limiting resistors or fuses to be fitted provided that they are made to an acceptable standard and the contact electrodes are shrouded.

Voltage indicating device

Test lamps and voltage indicators are recommended to be clearly marked with the maximum voltage which may be tested by the device and any short time rating for the device if applicable. This rating is the recommended maximum current that should pass through the device for a few seconds, as these devices are generally not designed to be connected for more than a few seconds.

Operating devices

It is not an adequate precaution to simply isolate a circuit electrically before commencing work on it. **It is vital to ensure that once a circuit or item of electrical equipment has been isolated, it cannot be inadvertently switched back on**.

A good method of providing full electrical and mechanical isolation is to lock off the device (or distribution board containing the device) with a padlock (see Figure 6.3), with the person working on the isolated equipment keeping the key in their pocket. A skilled person should be the only one allowed to carry out this or similar responsible tasks concerning the electrical installation wiring.

Figure 6.3 Device locked off with padlock

Many items of electrical equipment when installed are provided with a local means of switching or isolation. This is not only convenient for the user (as control is readily accessible) but it is a safety feature and indeed a requirement of BS 7671 *IEE Wiring Regulations* (Chapter 53) that 'equipment should be readily controlled, protected and isolated'. A good example of local isolation is when a motor is provided with a control unit as this can include the starter, switch and lock-off facility.

The types of equipment used in circuits to provide switching and isolation of the circuits, and indeed complete installations, can be categorised as having one or more of the following functions:

Figure 6.4 One-way lighting circuit with mains switch and breakers

- control
- isolation
- protection.

A simple example of the control, isolation and protection functions can be represented in any basic simple circuit. As an example, Figure 6.4 shows a one-way lighting circuit supplied from a distribution board having a mains switch and circuit breakers.

From Figure 6.4 we can see that:

- The distribution consumer unit combines all three functions of control, protection and isolation.

- The main double-pole switch can provide the means for switching off the supply and, when locked off, can provide complete isolation of the installation.

- The circuit breakers in the unit provide protection against faults and overcurrents in the final circuits and, when switched off and locked off, the circuit breakers can provide isolation of each individual circuit.

- The one-way switch has only one function, which is to control the circuit enabling the luminaire to be switched 'on' or 'off'.

On and off-load devices

Not all devices are designed to switch circuits 'on' or 'off'. It is important to know that when a current is flowing in a circuit, the operation of a switch (or disconnector) to break the circuit will result in a discharge of energy across the switch terminals.

You may well have had this happen when entering a dark room and switching on the light, where for an instant you may see a blue flash from behind the switch plate. This is actually the arcing of the current as it dissipates and makes contact across the switch terminals. A similar arcing takes place when circuits are switched off or when protective devices operate thus breaking fault current levels.

Fundamentally, an isolator is designed as an off-load device and is usually only operated after the supply has been made dead and there is no load current to break. An on-load device can be operated when current is normally flowing and is designed to make or break load current.

An example of an on-load device could be a circuit breaker, which is not only designed to make and break load current but has been designed to withstand make and break high levels of fault current. Remembering the three previous functions, it is important to install a device that meets the needs at a particular part of a circuit or installation. Some devices can meet the needs of all three functions. However, some devices may only be designed to meet a single function.

All portable appliances should be fitted with the simplest form of isolator, a fused plug. This, when unplugged from the socket outlet, provides complete isolation of the appliance from the supply.

Restoration of the supply

After the fault has been rectified, which may have resulted in either parts being replaced or simple reconnection of conductors, it is important that the circuit is tested for functionality. These tests may be simple manual rotation of a machine or the sequence of tests as prescribed in the *IEE Wiring Regulations*. For example, a simple continuity test will check resistance values, open and closed switches and their operation.

Remember

A plug is not designed to make or break load current and that the appliance should be switched off before removing the plug

General categories of fault

In this section you will recognise different types of faults and understand that consequences of the fault depend on its location within the installation or in a specific circuit. You should also understand the need for care when installing electrical systems and how poor, or lack of, maintenance could lead to such faults. The essential points to be covered are:

- position of faults (complete loss of supply at the origin of the installation and localised loss of supply)
- operation of overload and fault current devices
- transient voltages
- insulation failure
- plant, equipment or component failure
- faults caused by abuse, misuse and negligence
- prevention of faults by regular maintenance.

Position of faults

The knowledge of fault finding and the diagnosis of faults can never be complete because no two situations are exactly the same. Also, as technology advances and the systems we install are being constantly improved, then the faults that develop become more complicated to solve. Therefore an understanding of the electrical installation and the equipment we install becomes paramount.

To diagnose faults an electrician will adopt basic techniques and these can usually solve the most common faults that occur. However, there are occasions when it may be impractical to rectify a fault. This could be due to costs involved in down time, or the cost of repair being more than the cost of replacement, in which case it could mean the renewal of wiring and/or equipment or components. Such situations and outcomes must be monitored, and the client should be kept informed and be made party to the decision to repair or replace.

Almost all faults should never exist. A fault can be compared to an accident, where careful planning and thought can prevent accidents. It is therefore part of a designer's job to build into the design of the electrical installation fault protection and damage limitation, such as:

- Installing more than one circuit. When a fault occurs on one of the two circuits the fault can be limited to that circuit.
- Installing fuses and protective devices to disable the fault and limit its effect.
- Ensuring the ability to access, maintain and repair. Good maintenance will limit faults, and access to the installation allows for maintenance and repair.

Remember

Always keep the client informed

It is easier to find faults on installations where there are plenty of circuits. Indeed it is a requirement of the *IEE Wiring Regulations* BS 7671 Part 3 under **Installation circuit arrangements**, where Regulation 314-01-01 states 'Every installation shall be divided into circuits as necessary to:

i)   avoid danger and minimise inconvenience in the event of a fault

ii) facilitate safe operation, inspection, testing and maintenance.'

We can see that compliance with this Regulation will help us to locate faults more easily, as usually by a process of elimination (operating each fuse individually) or simply looking at the device to see which one has operated we can identify a defective circuit or device.

Protective devices and simple fuses are designed to operate when they detect large currents due to excess temperature. In the case of a short circuit fault, high levels of fault current can develop, causing high temperatures and breakdown of insulation. Such faults, if allowed to develop, can cause fires.

When considering the consequences of a fault, it is important to know that the location of the fault can limit its severity with regard to disruption and inconvenience

Meltdown of insulation due to overcurrent

Meltdown of plastic connector due to overcurrent

Figure 6.5 could be a typical layout of equipment for a small industrial or commercial installation where a three-phase supply is required and the load is divided on the three final consumer units, and will assist you in understanding simple installation layouts and the consequence of a located fault.

Key to Figure 6.5

A	Supply company's service with incoming supply cable and fuses.
A1	Supply company's cables feeding KWh meter.
B	Supply company's metering equipment KWh meter.
B1	Consumer's meter tails.
C	Consumer's main switch – typical TP&N 100 amp.
C1	Sub final circuit conductors brown phase.
C2	Sub final circuit conductors black phase.
C3	Sub final circuit conductors grey phase.
D	Consumer's unit with DP main switch and protective devices.
E	Consumer's unit with DP main switch and protective devices.
F	Consumer's unit with DP main switch and protective devices.
D1	Consumer's final circuits i.e. sockets and lighting.
E1	Consumer's final circuits i.e. sockets and lighting.
F1	Consumer's final circuits i.e. sockets and lighting.

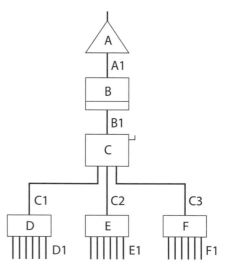

Figure 6.5 Typical three-phase supply for small installation

Complete loss of supply at the origin of the installation

With reference to Figure 6.5 it can be seen that a fault at points A to B1 could result in the complete loss of supply to the rest of the installation. The equipment at this point is mainly the property of the supply company, and faults that appear on this equipment are usually the result of problems with the supply, such as an underground cable being severed by workmen.

However, faults on the installation wiring would result in the supply company's protective devices operating but this is unusual and is only a consequence of poor design, misuse of the equipment or overloading of the supply company's equipment.

As an example of this, if the supply company's protective device or fuses were rated at 100 amperes per phase and the designer loaded each phase up at 200 amperes, this would result in a 100 per cent overload and the consequence of this would be the operation of the mains protective devices and the whole installation being without power.

Localised loss of supply

Looking again at Figure 6.5 we can recognise what happens when parts of the installation lose their supply. If a fault appeared on Cable C1, then the consumer unit (D) and all the final circuits from it would be dead. If a fault occurred on any one final circuit then only that circuit's protective device should have operated and only that circuit would be dead.

If a fault occurred on any one final circuit it should not affect any other final circuit or any other cable leading up to that consumer unit.

Because the system in Figure 6.5 is three phase, this should mean that a fault on any one phase should only result in that phase being affected, again limiting the consequence of a fault. This further highlights the reason for Regulation 314-01-01 that we mentioned earlier, which stated 'Every installation shall be divided into circuits as necessary to:

i) avoid danger and minimise inconvenience in the event of a fault

ii) facilitate safe operation, inspection, testing and maintenance.'

Operation of overload and fault current devices

When a fault is noticed, it is usually because a circuit or piece of equipment has stopped working and this is usually because the protective device has done its job and operated. The rating of a protective device should be greater than, or at least equal to, the rating of the circuit or equipment it is protecting, e.g. 10 x 100 watt lamps equate to a total current use of 4.35 amperes. Therefore a device rated at 5 or 6 amperes could protect this circuit.

A portable domestic appliance which has a label rating of 2.7 kW equates to a total current of 11.74 amperes. Therefore a fuse rated at 13 amperes should be fitted in the plug.

Definitions

Overload current – an overcurrent occurring in a circuit which is electrically sound

Overcurrent – a current exceeding the rated value

Earth fault current – a fault current which flows to earth

Short circuit current – an overcurrent resulting from a fault of negligible impedance between live conductors

Protective devices are designed to operate when an excess of current (greater than the design current of the circuit) passes through it. The fault current's excess heat can cause a fuse element to rupture or the device mechanism to trip, dependent on which type of device is installed. These currents may not necessarily be circuit faults, but short-lived overloads specific to a piece of equipment or outlet. The Regulations categorise these as **overload current** and **overcurrent**.

For conductors, the rated value is the current-carrying capacity. Most excess currents are, however, due to faults, either **earth faults** or **short circuit** type which cause excessive currents.

Whichever type of fault occurs the designer should take account of its effect on the installation wiring and choose a suitable device to disconnect the fault quickly and safely. The fundamental effect of any fault is a rise in current and therefore a rise in temperature.

High temperature destroys the properties of insulation, which in itself could lead to a short circuit. High currents damage equipment, and earth fault currents are dangerous to the body from electric shock. Some examples of each are given below.

Overload faults

- adaptors used in socket outlets exceeding the rated load of circuit
- extra load being added to an existing circuit or installation
- not accounting for starting current on a motor circuit.

Short circuit faults

- insulation breakdown
- severing of live circuit conductors
- wrong termination of conductors energised before being tested.

Wrong type of starter in fluorescent tube has melted due to excess power demand

Earth faults

- insulation breakdown
- incorrect polarity
- poor termination of conductors.

Fuse holder has melted due to overloading

Poor termination of (cpc) circuit protective conductors

Transient voltages

A transient voltage can be defined as a variance or disturbance to the normal voltage state, the normal voltage state being the voltage band that the equipment is designed to operate.

It is the installation designer's responsibility to stipulate adequate sizes for the circuit conductors. But this sizing procedure will only prevent voltage drop during normal circuit conditions. Designers cannot prevent transient voltages as they are outside their control, but they can compensate by including equipment that will protect against such voltage variations and disturbances.

It is now becoming common practice to install filter systems to IT circuits, which provide stabilised voltage levels. These devices can also reduce the effect of voltage spikes by suppressing them. If transient voltages are not recognised they can cause damage to equipment and lead to a loss of data that could prove costly, especially if the user was an organisation such as a bank.

Some common causes of transient voltages are:

- supply company faults
- electronic equipment
- heavy current switching (causing voltage drops)
- earth-fault voltages
- lightning strikes.

Insulation failure

Insulation is designed primarily to separate conductors and to ensure their integrity throughout their life. Insulation is also used to protect the consumer against electric shock by protecting against direct contact and is also often used as a secondary protection against light mechanical damage, such as in the case of PVC/PVC twin and cpc cables. However, as previously mentioned, the breakdown of electrical insulation can lead to short circuit or earth-faults.

When insulation of conductors and cables fail, it is usually due to one or more of the following:

- poor installation methods
- poor maintenance
- excessive ambient temperatures
- high fault current levels
- damage by third party.

Did you know?

Transient voltages are becoming increasingly problematic with the expanding use of IT equipment, not only in commerce and industry, but also in the home

Did you know?

If transient voltages are thought to exist on installations, monitoring equipment can be installed to record events over periods of time. The results of the monitoring can usually trace the cause

Poorly connected 13 A plug top resulting in dangerous working conditions

Plant, equipment and component failure

It is said that nothing lasts forever and this is certainly true of electrical equipment. There will be some faults that you will attend that will be the result of a breakdown simply caused by wear and tear, although it must be said that planned maintenance systems and regular testing and inspections can extend the life of equipment. Some common failures on installations and plant are:

- switches not operating – due to age
- motors not running – new brushes required
- lighting not working – lamp's life expired
- fluorescent luminaire not working – new lamp or starter needed
- outside PIR not switching – ingress of water causing failure
- corridor socket outlet not working due to poor contacts created by excessive use/age.

Burnt out starter

Poorly wired socket outlet

Faults caused by misuse, abuse and negligence

A common reason for faults on any electrical system or equipment is misuse, where the system or the equipment is simply not being used as it was designed. Every item of electrical equipment comes with user instructions that usually cover procedures and precautions, and such instructions should be read carefully as misuse can lead to invalidation of the guarantee.

When an electrical installation is completed, a manual is normally handed over to the client that includes all product data and installation schedules and test results. This data will help the client when additions to the installation are made, inspections and tests are carried out or to assist in maintaining the installation.

Some faults are caused by carelessness during the installation process, and simple faults caused by poor termination and stripping of conductors can then lead to serious short circuits or overheating. Therefore good mechanical and electrical processes should always be carried out and every installation should be tested and inspected prior to being energised. Some examples of faults arising from misuse are given below.

User misuse

- using an MCB as a switch where constant use could lead to breakdown
- unplugging on-load portable appliances thus damaging socket terminals
- damp accessories, for instance, due to hosing walls in dry areas
- MCBs' nuisance-tripping caused by connecting extra load to circuits.

Remember

If you are called out to repair/rectify a fault and you find that misuse of the system has led to the fault, then the guarantee for the installation may be invalidated by the fault. This may lead to the client having to pay for the repair costs

MCB

Installer misuse

- poor termination of conductors – overheating due to poor electrical contact
- loose bushes and couplings – no earth continuity – electric shock risk!
- wrong size conductors used – excessive voltage drop – excess current which could lead to an inefficient circuit and overheating of conductors
- not protecting cables when drawing them in to enclosures, causing damage to insulation
- overloading conduit and trunking capacities, causing overheating and insulation breakdown.

Poor termination of conductors causing electric shock risk

Prevention of faults by regular maintenance

A well designed electrical installation should serve the consumer efficiently and safely for many years, whereas an installation that isn't regularly inspected and tested could lead to potential faults and areas of weakness not being discovered. Even a visual inspection can uncover simple faults which, if left, could lead to major problems in the future.

Large organisations in the commercial sector and industry have realised the need for regular inspections and planned maintenance schedules. Some major supermarkets choose to have regular maintenance take place when the stores are closed to the public, as they realise a store closure or part closure to carry out corrective electrical work could cost sales. A typical example of maintenance in shops is lamp cleaning, changing and repairs and servicing to refrigeration and air conditioning. Large manufacturing companies usually carry out maintenance during planned holiday periods when the workforce is on annual leave and machinery, equipment and the electrical installation wiring can be accessed without loss of production.

Carrying out repairs and servicing in this way gives electricians and engineers the opportunity to test installations, carry out remedial work and to do general servicing on equipment and machinery.

Specific types of fault

The type of equipment we install as electricians is normally chosen by a designer in line with the needs of the client's specification. We therefore often take it for granted that the equipment and wiring we fit is right for the job, and the designer has not only met the needs of the client but also the environment.

The manufacturer and the types of cables and accessories that you install must comply with BS and BSEN Standards and must be installed in accordance with BS 7671 *IEE Wiring Regulations*. Chapter 52 of the Regulations covers Selection and Erection of Equipment, in particular:

- selection of type of wiring system
- external influences
- current carrying capacity of conductors
- cross-sectional area of conductors
- voltage drop
- electrical connections
- minimising spread of fire
- proximity to other services
- selection and erection in relation to maintainability.

The items listed in this area of the Regulations should have been accounted for in the initial design but it is the responsibility of the installer to enforce them. This is achieved by adopting good practice when installing, and understanding the consequences (faults) that would occur if such good practice and Chapter 52 were ignored.

Cable interconnections

Cable interconnections are used on many occasions in electrical installations. Their use is generally seen to be poor planning, and good design would limit their use. However, they are used in one or more of the following ways:

- lighting circuit joint boxes for phase, neutral, switch wire and strappers
- power circuits, ring final socket outlet wiring and spurs etc.
- street lighting and underground cables where long runs are needed
- general alterations and extensions to circuits when remedial work is being done
- rectification of faults or damage to wiring.

Where they must be used, they should be mechanically and electrically suitable. They must also be accessible for inspection as laid down by Regulation 526-04-01, which states that 'every connection and joint shall be accessible for inspection, testing and maintenance'. There is an exception to this Regulation when any of the following is used:

- a compound or encapsulated joint (commonly underground)
- a connection between a cold tail and heating element (under floor)
- a joint made by welding, soldering, brazing or compression tool.

The joints in non-flexible cables shall be made by soldering, brazing, welding, mechanical clamps or of a compression type. The devices used should relate to the size of the cable and be insulated to the voltage of the system used. Connectors used must be of the appropriate British Standard and the temperature of the environment must be considered when choosing the connector.

Where cables having insulation of dissimilar characteristics are to be jointed, for example XLPE thermosetting with general purpose PVC, the insulation of the joint must be to the highest temperature of the two insulators. The most common types of terminating devices used are:

- plastic connectors
- porcelain connectors
- soldered joints
- Screwits
- uninsulated connectors
- compression joints
- junction boxes.

Screwit   RB4 Junction box

Cable interconnections are usually seen as the first point to investigate in the event of a fault. This is because they are seen as the weak link in the wiring system and because they are usually readily accessible.

The mechanical and electrical connection when joining two conductors together relies on good practice by the installer to ensure that the connection is made soundly. The terminating device used should:

- be the correct size for the cross-section of the cable conductor
- be at least the same current rating as the circuit conductor
- have the same temperature rating as the circuit conductor
- be suitable for the environment.

The most common fault at this point would probably be due to a poor/loose connection, which would produce a resistive joint and excessive heat that could lead to insulation breakdown or eventually fire. When a fault occurs at a cable interconnection it may not necessarily be due to the production of heat but to the lack of support or strain being placed on the conductors, which may lead to the same outcome as above.

Remember

Follow the safe isolation procedures before removing any cover or lid from any termination point

Cable terminations, seals and glands

Terminations

The same rules as for cable interconnections apply to the termination of cables, i.e. good mechanical and electrical connections. The same consequences will occur if they are not adhered to, namely:

- Care must be taken not to damage the wires.

- BS 7671 requires that a cable termination of any kind should securely anchor all the wires of the conductor that may impose any appreciable mechanical stress on the terminal or socket.

- A termination under mechanical stress is liable to disconnect, overheat or spark.

- When current is flowing in a conductor a certain amount of heat is developed and the consequent expansion and contraction may be sufficient to allow a conductor under stress, particularly tension, to be pulled out of the terminal or socket.

- If the PF improvement capacitor became disconnected in a luminaire, the circuit current would increase.

- A fault caused at a poorly connected terminal would be known as a high resistance fault.

- One or more strands or wires left out of the terminal or socket will reduce the effective cross-sectional area of the conductor at that point. This may result in increased resistance and probably overheating.

- Poorly terminated conductors in circuits that continue to operate correctly are commonly known as latent defects.

Types of terminals

There are a wide variety of conductor terminations. Typical methods of securing conductors in accessories include pillar terminals, screw heads, nuts and washers, and connectors as shown in Figures 6.6, 6.7 and 6.8.

Figure 6.6 Pillar terminal

Figure 6.7 Screwhead, nut and washer terminals

Figure 6.8 Strip connectors

Seals and entries

Where a cable or conductor enters an accessory or piece of equipment, the integrity of the conductor's insulation and sheath, earth protection and the enclosure or accessory should be maintained. Some cables and wiring systems have integrated mechanical protection i.e.:

- PVC/PVC twin and cpc
- PVC/steel wire armoured
- PVC covered mineral insulated copper cable (MICV)
- FP200.

FP200 cable

Twin and cpc PVC insulated and sheathed cable

MICV cable with pot and seal

When these and similar cables are installed, their design capabilities should not be degraded and special glands and seals, which are produced by the makers, should be used. Where PVC/PVC cables enter accessories, the accessory itself should have no loss of integrity and there should be no damage to any part of the cable.

When carrying out a visual inspection of an installation, either at the completion of works or at a periodic inspection, the checking of terminations of cables and conductors is an integral part of the inspection. Some items listed for checking in Regulation 721-01-01 include:

- connection of conductors
- identification of conductors
- routing of cables
- selection of conductors in accordance with the design
- correct connection of accessories and equipment.

These checks are made to comply with Regulation 712-01-02(a), which states 'The inspection shall be made to verify that the installed electrical equipment is not visibly damaged or defective so as to impair safety'.

We should also routinely check:

- for correct entry of cable into the accessory
- that the correct type of gland and seal has been used
- that accessory or enclosure seals and building structure/cable routes sealed.

Accessories including switches, control equipment, contactors, electronic and solid state devices

Faults can appear on most items of equipment that we install. The most common fault is due to wear of the terminal contacts, which constantly make and break during normal operation. All items of equipment should be to BS or BSEN and must be type tested by the manufacturer.

Some accessories break down due to excessive use, and through experience the electrician can usually recognise them. During regular inspections and maintenance of an installation the inspector will check 10 per cent of accessories visually, which will usually be items that are constantly being used such as:

- entrance hall switches
- socket outlets in kitchens for kettles
- cleaner's socket
- any item which is regularly being switched on or off.

Control equipment

This is also termed switchgear and is found at the origin of the supply after the metering equipment. It can take many forms; a few examples are:

- domestic installation – double pole switch in consumer unit
- industrial installation – single-phase double pole switch fuse or three-phase triple pole and neutral switch fuse
- large industrial installation moulded case circuit breakers – polyphase
- switch rooms industrial/commercial – as above, also oil and air blast types.

The switchgear is categorised not only in normal current ratings for load in amperes, but also for fault current ratings in 1000 amperes or kA. If we refer back to Figure 6.5 on page 194, item C is a typical example of switchgear control equipment.

If a fault occurred on the outgoing conductors from these devices then the level of fault current would be dependent on the location. The basic rule for fault current level is that the nearer to the supply that the fault occurs, then the greater the level of fault current. This is due to there being less impedance (resistance) to impede the current flow at that point, whereas the further away from the origin of the supply the greater the impedance (resistance) to impede the current flow.

Remember

Operating mechanisms do not normally break down and where contact points can be serviced and the build up of carbon removed, this will extend the life of the accessory. However, many items are not serviceable and when a breakdown occurs through constant use, a repair is neither a choice nor cost-effective

The operation of this type of control gear, either manually or automatically, due to a fault would leave all circuits fed from such equipment dead. Manual operation of such devices is usually for maintenance reasons and it should be noted that permission of the client by prior arrangement would need to be obtained before commencing such work.

Protective devices are usually an integral part of switchgear, which will be one of the following:

- high breaking capacity fuses HBC or HRC
- moulded case circuit breakers MCCB.

Specialist information and manufacturer's instructions are needed for replacing fuses that have operated and when re-setting tripped circuit breakers.

In older type installations you may come across switchgear that is prone to fault due to lack of maintenance. For example oil type circuit breakers have their contact breaker points submerged in a special mineral-based oil. The oil, which quenches arcing, assists in breaking high levels of load and fault currents. If the oil viscosity and level is not regularly checked, this could lead to high levels of fault current circulating, leading to equipment damage and insulation breakdown or even fire.

Contactors

Contactors are commonly used in electrical installations. They can be found in:

- motor control circuits
- control panels
- electronic controllers
- remote switching.

The basic principle of operation of a contactor is the use of the solenoid effect, in which a magnetic coil is used in the energised or de-energised mode and spring loaded contacts for auxiliary circuits are either made or broken when the coil has been energised or de-energised.

Operating coils used in motor control circuits are energised when the start button is operated, with the most common fault associated with their use being coils burning out. This can be due to age, but can be due to the wrong voltage type being used, i.e. a 230 volt coil being used across a 400 volt supply.

Remote switching can utilise the contactor with great effect and convenience because the contacts of large, load switching contactors can be used to carry large loads, i.e. distribution boards and large lighting loads such as used in car parks, floodlights etc. The benefit of the system is that a local 5 A switch in an office or reception area can be operated when entering or leaving, thus energising the coil of the contactor and allowing heavy loads to be carried through the contactor's contacts.

Moulded case circuit breaker

Did you know?

As contacts are regularly being made and broken, a build-up of carbon appears on the contact points. On larger contactors these can simply be cleaned periodically using fine emery paper to remove the residue and make the electrical contact point more efficient

Remember

On smaller contactors used in electronic circuits it is not possible to clean contact points, and when breakdown occurs it is normal practice to replace rather than repair

Electronic and solid state devices

Solid state and electronic equipment work within sensitive voltage and current ranges in millivolts and milliamperes, and are consequently sensitive to mains voltage and heat.

In the case of resistors on circuit boards, excessive heat produces more resistance and eventually an open circuit when the resistor breaks down. We can use a low-resistance ohmmeter to check resistor values.

In the case of capacitors, voltages in excess of their working voltage will cause them to break down, which will result in a short circuit. Such equipment obviously requires specialist knowledge, but some basic actions by the electrician when carrying out tests and inspections can prevent damage to this type of equipment. When testing, such equipment should be disconnected prior to applying tests to the circuits. The reason for this is that some test voltages can damage these sensitive components and their inclusion in the circuit would also give an inaccurate reading.

We have discussed transient voltages and these are a major cause of faults on components and equipment of this type. These voltages can arise from supply company variations or lightning strikes, and therefore most large companies protect their equipment from such voltages and employ specialist companies to install lightning protection and filtering equipment.

Instrumentation

Switch room panels have integrated metering equipment. In this part of the installation please note that **large currents are being monitored**. This monitoring allows the consumer to view current and voltage values at particular times of the day. Some panels will have equipment which monitors and records these values, allowing the consumer to plot peak energy times and budget for energy costs.

As these instruments measure large currents, transformers are used to reduce current and voltage values, and this reduces the size of instrument needed. These transformers are known as CTs and VTs. However, great care must be taken when servicing and repairing such equipment as large voltages can exist across the transformer terminals. The common remedial work on such equipment is the replacement of faulty instruments and burnt-out transformers. It is essential that safe isolation procedures are followed.

Protective devices

These devices operate in the event of a fault occurring on the circuit or equipment that the device is protecting. Typical faults include:

- short circuits between live conductors – phase to neutral single-phase, phase to phase three-phase
- earth faults – between any live conductor and earth
- overcurrents.

The most common reason for a device not operating is that the wrong type or rating of device has been used. Therefore, when replacing or re-setting a device after a fault has occurred, it is important to replace the device or repair it with the same type and rating. This should only be done after the fault has been investigated and corrected. The consequence of switching on a device with an un-repaired fault is for the device to operate again. This type of activity will cause damage to the circuit wiring, the equipment and possibly the device.

Luminaires

The most common fault with luminaires is expiry of lamp life, which obviously only requires replacement. Discharge lighting systems employ control gear which on failure will need replacement as they cannot be repaired. Discharge type lighting may have problems with the control circuit.

Common items of equipment in the luminaire control circuit are:

- The capacitor used for power factor correction. If this had broken down it would not stop the lamp operating but would prevent the luminaire operating efficiently.

- The choke or ballast used to create high p.d. to assist in the lamp starting. A common item which could break down and needs replacement.

- The starter, which is used to assist the discharge across the lamp when switched on at the start. This is the usual part to replace when the lamp fails to light.

Flexible cables and cords

This type of conductor is used to connect many items within an electrical installation, e.g.:

- pendant ceiling rose lamp holders

- flex outlet spur units

- fused plugs to portable appliances

- immersion heaters

- flexible connection to fans and motors.

Regulation 521-01-01 states that all flexible cables and cords shall comply with the appropriate British and Harmonised Standard. The most common faults that occur with this type of conductor are likely to arise from poor choice and suitability for the equipment and the environment. Common faults relating to flexible cables and cords are:

- poor terminations into accessory – conductors showing etc.

- wrong type installed i.e. PVC instead of increased temperature type

- incorrect size of conductor – usually too small for load

- incorrectly installed when load bearing in luminaires.

Such problems should be identified during the visual inspection stage.

Did you know?

When a protective device has operated correctly and this device was the nearest device to the fault, **discrimination** is said to have occurred

Did you know?

Many fluorescent luminaires have starter-less electronic control gear, which is not only quick-start with increased efficiency, but requires less maintenance. They have a longer lamp life but when the quick-start unit fails it will need replacement

Portable appliances and equipment

Flexible cables and cords are used to connect many items of fixed and portable equipment to the installation wiring, and most faults on cords and flexible cables usually relate to poor choice or installation. Portable appliances come pre-wired with a fused plug to BS 1363.

When supplying equipment with flexible cables and cords it is important that the requirements of the *IEE Wiring Regulations* are followed but specifically:

- correct size conductor used for the load
- correct type used for the environment – temperature – moisture – corrosion
- correct termination method used to ensure good connection and no stress
- correct type and rating of protective device used to protect the cable and appliance.

Knowledge and understanding of electrical systems and equipment

Experience and understanding of electrical installations takes years to learn and a vast amount of knowledge is gathered during your apprenticeship. However, putting that knowledge into practice only really occurs when you take up responsibilities as an electrician.

When analysing faults on electrical installations, even the trained electrician will need to ask questions to help in rectifying faults.

Hopefully by now you should have come across many different types of wiring systems, equipment, enclosures and protective devices. Equally, you should have now used the procedures for safe isolation and locking off procedures many times. However, when a fault occurs, the art of finding information and asking questions can be a daunting task.

Understanding the electrical system, installation and equipment

These can generally be categorised as follows:

- voltage – 230 volt single-phase or 400 volt three-phase
- installation type – domestic, industrial or commercial etc.
- system type – lighting, power, fire alarm system, emergency lighting, heating etc.

If a fault occurs on a system which is in the process of being installed then the information and data for the system should be at hand. However, the big problem starts when you are asked to rectify a fault on a system that you have had no experience or knowledge of. In this situation you will need to access drawings and data to familiarise yourself with layout, distribution boards etc. The type of information that you will require when called out to a fault can be listed as:

- type of supply – single phase or three phase
- nominal voltage – 230 volt or 400 volt
- type of earthing supply system – TT, TN-S, TN-C-S
- type of earthing arrangement – EEBAD
- types of protective device – HRC, MCB, RCD etc.
- ratings of devices
- location of incoming supply services
- location of electrical services
- distribution board schedules
- location drawings
- design and manufacturer's data.

Remember

An electrician's knowledge and training can be a great advantage in recognising and tracing faults, but the task will be made easier and the fault will be remedied quicker if the information listed is available. As a first choice, where this information is not readily available you should either seek advice or try to obtain it

Optimum use of personal and other person's experience of systems and equipment

Your own knowledge of an installation can be an asset not only in the fault finding process, but also to your company for future business. If your company can problem-solve quickly and be relied upon by a client when a problem occurs, it will be your company that will be the first to be called.

The fault may not exist on the installation wiring but on auxiliary equipment such as refrigeration and air conditioning. Unless you have specialist-trained knowledge of such equipment, it is not safe to attempt to rectify the faults and your company may not be insured to work on such systems. However, if the fault is on the circuit supplying these systems, the fault may be investigated and rectified. You may also have to work alongside other specialists, assisting each other in commissioning and testing such systems. On no account should you attempt to investigate or rectify faults on their systems without knowledge or experience and training.

Equally, you will often be asked to repair faults on wiring and equipment that have been caused by inexperienced personnel attempting to install or repair circuits or equipment.

When you first arrive at any installation to investigate and rectify a fault, you should always use the personnel present to help you to obtain any background to the fault and relevant information, and if possible seek the specialist knowledge of the person responsible for the electrical installation or general maintenance.

The different people that you will be dealing with to obtain special and essential knowledge and information could be:

- the electrician who may have previously worked on the system
- design engineer
- works engineer
- shift engineer
- maintenance electrician
- machine operator
- home owner
- site foreman
- shop manager
- school caretaker.

These persons may have access to:

- operating manuals
- wiring and connection diagrams
- manufacturer's product data/ information
- maintenance records
- inspection and test results
- installation specifications
- drawings
- design data
- site diary.

Their experience of the installation and day-to-day knowledge is essential. It will help you to solve, replace or repair the fault more efficiently.

Stages of diagnosis and rectification

To be able to competently investigate, diagnose and find faults on electrical installations and equipment is one of the most difficult jobs for an electrician and therefore, if the electrician is to be successful in fault finding and diagnosis, a thorough working knowledge of the installation or equipment involved is essential.

Consequently, the person carrying out the procedure should take a reasoned approach and apply logical steps or stages to the investigation and subsequent remedy. The electrician should also realise their knowledge limitations and seek expert advice and support where necessary.

In an ideal world an electrician should not embark on any testing or fault finding without some forward planning. However, an emergency or dangerous fault, because of its very nature, may allow very little time for planning the remedy or repair. That said, some faults that are visible and straightforward can be easily repaired, and some careful planning and liaison with the client will limit disruption.

Inspection and testing prior to energising a new installation can be important and can save embarrassment by rectifying problems before circuits are energised. As an example, pre-energising tests can be extremely important when checking the function of switching circuits, particularly two-way and intermediate switching.

The electrician should have knowledge of inspection and testing in order to fault find competently – in particular the correct use of test instruments, the choice of instrument for testing, and knowledge of each instrument's range and limitations.

Logical stages of diagnosis and rectification

Some faults can be rectified very easily, especially when the electrician has a working knowledge of the installation. But there are many occasions when you may be called out to a repair and there is no information for you. It is on these occasions that your years of training and knowledge of wiring systems have to be used. However, this knowledge is not always enough and a sequential and logical approach to rectifying a fault and gathering information is needed. Such an approach could be:

- **Identify the symptom**: This can be done by establishing the events that led up to the problem or fault on the installation or equipment.

- **Gather information**: This is achieved by talking to people and obtaining and looking at any available information. Such information could include manufacturer's data, circuit diagrams, drawings, design data, distribution board schedules and previous test results and certification.

- **Analyse the evidence**: Carry out a visual inspection of the location of the fault and cross-reference with the available information. Interpret the collected information and decide what action or tests need to be carried out. Then determine the remedy.

Remember

The *IEE Wiring Regulations* on testing can be a useful guide, but the *Electricity at Work Regulations* and safe use of test equipment should be followed during the process

Remember

Information is a necessity and no fault finding should commence without the relevant information and background to the fault being made available to the person carrying out the work. Some faults can be recognisable by simply analysing test results and on further investigation can be remedied simply

- **Check supply:** Confirm supply status at origin and locally. Confirm circuit or equipment when fault is isolated from supply.

- **Check protective devices:** Check status of protective devices. If they have operated this would determine location of fault on circuit or equipment.

- **Isolation and test:** Confirm isolation prior to carrying out sequence of tests.

- **Interpret information and test results:** By interpreting the test results, the status of protective devices and other information, the fault may be identified and remedied/rectified.

- **Rectify the fault:** This may be done quite simply or parts or replacement may be needed.

- **Carry out functional tests:** Before restoring the supply, the circuit or equipment will need to be tested not only electrically but also for functionality.

- **Restore the supply:** Care must be taken that the device has been reset or repaired to correct current rating. Make sure the circuit or equipment is switched off locally before restoring supply.

In summary, the logical sequence of events would be:

1. Identify the symptoms

2. Gather information

3. Analyse the evidence

4. Check supply

5. Check protective devices

6. Isolation and test

7. Interpret information and test results

8. Rectify the fault

9. Carry out functional tests

10. Restore the supply.

Information required for fault location

We have already said that some information regarding the fault will be gathered by asking people questions relating to the events leading up to the fault. If the fault occurred on the installation when the electrician was working on it then the information will be to hand. Most often the situation regarding faults is when we are called out to a breakdown and asked to carry out repairs. It is when we are called out to a fault or repair that we will need information such as:

- operating manuals

- wiring and connection diagrams

- manufacturer's product data/information
- maintenance records
- inspection and test results
- installation specifications
- drawings
- design data
- site diary.

Interpretation of test results

Some faults can be recognised at the installation stage when the testing and inspection process is carried out. We should be able to recognise typical test results for each non-live test and interpret the type of fault that may exist.

Non-live tests

The *IEE Wiring Regulations* BS 7671 Part 7 – Chapter 71 lists non-live and live tests. The non-live tests are carried out on the installation wiring circuits before they are ever energised. These tests will confirm the integrity of the circuit and they can be listed as:

- continuity of protective conductors
- continuity of ring final circuit conductors
- insulation resistance
- polarity.

Continuity of protective conductors

Protective conductors can be found in final circuit wiring and from the main earthing terminals to all metalwork that is to be found in the installation. It is important to verify the continuity of such conductors because we rely upon them for safety in the event of an earth fault and the operation of protective devices and fuses. A low-resistance ohmmeter would be used on conductors to verify a low resistance value. The reading obtained will depend on the length and area of the conductor, as resistance is proportional to length and inversely proportional to cross-sectional area.

If you are testing for the continuity of equipotential bonding the test current should be a minimum of 200 mA (Regulation 713-02-01). A typical reading for earth continuity would be between 0.01 to 0.05 ohms. If a reading greater than 0.5 ohms was obtained, although on the upper side of acceptable, this may need investigating. Although it may be acceptable in relation to the conductor type and size, terminations may require checking for effectiveness.

Continuity of ring final circuit conductors

This test, carried out on the circuit wiring, is designed to highlight open circuits and interconnections within the wiring, and a low-resistance ohmmeter is used to carry out this test. Typical readings will be 0.01 to 0.1 ohms dependent on the conductor size and length of circuit wiring. If the circuit is wired correctly, then the readings on the instrument will be the same at each point of test. If variable readings are found at each point of the test, this may indicate an open circuit or interconnections. If these types of fault exist, the consequence can be an overload on part of the circuit wiring.

Insulation resistance

This test is designed to confirm the integrity of the insulation resistance of all live conductors, between each other and earth, and the type of instrument is known as an insulation resistance tester. Typical test results would be around 50 to 100 megohms, i.e. 50 to 100 million ohms.

Table 71A of the *IEE Wiring Regulations* indicates minimum values between one quarter of a million and one million ohms. These are minimum values and, in practice if these values existed, the circuit may need further investigation. Testing should be carried out with a test appliance capable of supplying the test voltage indicated in Table 71A when loaded with 1 mA. Such minimum values exist because the test voltage is divided by the delivered value of current of the instrument, which will indicate the actual insulation resistance of the test:

$$\frac{250 \text{ volts}}{1 \text{ mA}} = 0.25 \text{ megohms}$$

Depending on where the test takes place, variable values of resistance may be recorded. In the case of a test on the supply side of a distribution board (Figure 6.9), a group value reading will indicate circuits connected in parallel (Figure 6.10).

Figure 6.9 Test on supply side of distribution board

Figure 6.10 Test on circuits connected in parallel

From Figures 6.9 and 6.10, it can be seen that when testing grouped circuit conductors, a parallel reading would indicate a poor reading, but if each circuit were tested individually then the actual reading would be acceptable.

Polarity

The polarity test can be carried out using a low resistance ohmmeter as used for typical continuity testing. This polarity can be confirmed during the continuity test on the circuit and protective conductors. Typical test results will depend on the resistance of the conductors under test but will generally be low, around 0.01 to 0.1 ohms. Values in excess of this may indicate an open circuit or incorrect polarity.

Functional testing

This is basically mentioned in Regulation 713-12-01 of the *IEE Wiring Regulations*, which states that where a residual current device provides indirect contact protection and supplementary protection against direct contact, then the effectiveness of the device should be confirmed by a test simulating typical fault-current conditions.

Such a test would use an appropriate RCD test instrument and the test should confirm the operation of the device independently of the device's integral test button. Residual current circuit breakers or devices are rated in milliamperes and the test should show that the device operates within the milliampere ratings of the device and within the time constraints. Typically if the rated current of the device was 30 mA then the instrument should prove operation at this value within 200 milliseconds.

Regulation 713-12-02 states that assemblies such as switchgear and control gear, drives, controls and interlocks shall be subjected to a functional test, to show that they are properly mounted, adjusted and installed in accordance with relevant requirements of the *IEE Wiring Regulations*. Typical functional tests should be applied to:

- lighting controllers, switches etc.
- motors, fixings, drives, pulleys etc.
- motor controllers
- controls and interlocks
- main switches
- isolators.

Earth-fault loop impedance testing

Overcurrent protective devices must, under earth-fault conditions, disconnect fast enough to reduce the risk of electric shock. This can be achieved if the actual value of the earth loop impedance does not exceed the tabulated values given in BS 7671. The purpose of the test, therefore, is to determine the actual value of the loop impedance (Z_s) for comparison with those values.

The test procedure requires that all main equipotential bonding is in place. The test instrument is then connected by the use of 'flying' leads to the phase, neutral and earth terminals at the remote end of the circuit under test. Press to test and record the results.

Once Z_s and the voltage have been established at the remote point in the circuit, you can divide the voltage by Z_s to give you the fault current to earth. Apply this calculation to the time current characteristic graphs shown in BS 7671 and you will be able to determine the actual disconnection time of the circuit.

The limitation and range of instruments

The correct choice of instrument for each particular test should now be known, but for those who are still unsure refer to Table 6.1.

Test	Instrument	Range
Continuity of protective conductors	Low resistance ohmmeter	0.005 to 2 ohms or less
Continuity of ring final circuit conductors	Low resistance ohmmeter	0.05 to 0.8 ohms
Insulation resistance	High reading ohmmeter	1 megohm to greater than 200 megohms
Polarity	Ohmmeter	Low resistance
	Bell/Buzzer	None

Table 6.1 Limitation and range of test instruments

Some instruments can provide more than one facility and many manufacturers provide such instruments. Inspectors and testers should know the range of operation of their instruments, be conversant with their operation and understand the instructions on the instrument.

When carrying out testing procedures, and before any instrument is used, some simple checks should be carried out to ensure that the instrument is operating within its range and calibration. A suitable checklist for use prior to testing is given below.

- ☑ Check the instrument and leads for damage or defects.
- ☑ Zero the instrument.
- ☑ Check the battery level.
- ☑ Select the correct scale for testing. If in doubt, ask or select the highest range available.
- ☑ Check the calibration date and record the serial no.
- ☑ Check leads for open and closed circuit prior to test.
- ☑ Record test results.
- ☑ After test, leave selector switches in off position. Some analogue instruments turn off automatically to save battery life.
- ☑ Always store instruments in their cases in secure, dry locations when not in use.

Factors affecting the fault repair process

Having achieved the aim of locating the fault it may be a simple decision by the electrician to eliminate the cause and re-energise the circuit for the equipment or circuit to function again. However, in many cases there are issues that will not allow the power to be switched on to the circuit or equipment under scrutiny. This judgement in many cases cannot be made by the electrician and may need to be discussed by third parties i.e. the client, the manufacturer and the person ordering the work. In this section we will look at the factors which will influence this decision-making process as to whether the equipment or circuit is replaced or repaired and whether the fault can be remedied instantly or at a pre-determined future time.

Cost of replacement

When successfully finding a fault it would not make for good customer relations to go ahead and put the fault right without first considering the consequences to your company or your client. There are points that need consideration such as: 'What will the cost of replacement be?'

This question can only be answered by liaison and consultation with your engineer and the customer, as an agreement must be reached on how to proceed. It may well be that some electrical contractors have an agreement or service contract in place where remedial works are carried out on a day work and contracted basis. This is where a prior contract agreement is signed and the contractor carries out maintenance and repair work at pre-determined rates. However, where this is not the case the question may arise: 'Will the replacement cost be exorbitant?'

It could be that the price quoted for replacement is high, which may result in the customer using your company's quotation to compare with other companies' prices. It may lead to negotiations where only part of a faulty piece of equipment is replaced or repaired, which may reduce the outlay costs to the customer. It is often the case that the customer is given a choice of options to help them decide what to do; they will compare these options and make the decision.

The options may be:

● full replacement

● part replacement

● full repair.

In offering such options it can be seen generally that the time taken to carry out a repair or part replacement may in many cases outweigh the time and cost of a full replacement.

Availability of replacement

As we have said before, in many cases where a complete replacement of equipment or machinery is needed, it is because the fault has arisen due to the life span of the equipment being exceeded. In these cases, the equipment or machinery may have been in operation for so long that a repair may not be possible and the item may even be obsolete. When this happens, a suitable replacement must be found.

The most obvious decision to be made is that of finding a replacement that closely matches the original specification of the faulty item. But this is not always possible and the factor that, in reality, will affect the selection process becomes the availability of the equipment. This is important and will influence the decision to order. It is often the manufacturer who has the replacement item in stock and has a fast delivery that will be given the order to supply, as this will enable the electrician to remedy the fault quickly.

Downtime under fault conditions

It is often the case that the consequence of a fault on a circuit or item of equipment will be costing the client money. This can be due to:

- lost production where manpower and machinery are not working
- lost business where products or services are not being provided to the customer
- data loss because of power failure or disruption during repair.

Sometimes the fault may not be on the electrical installation but on the supply to the consumer's premises. In such cases the rectification of the fault is out of your company's responsibility and control. However, in such circumstances and where the loss of supply is prolonged, it may make good business sense, where possible, to provide a temporary supply to the customer.

Availability of resources and staff

We have already mentioned the availability of equipment due to choice, manufacturer and delivery time, but the availability of staff is often the cause of any delay in rectifying faults and carrying out repairs.

Some faults and repairs may be minor and need little resource in terms of materials and labour. The response time for small faults could be as low as minutes and hours, but in the case of major industrial and commercial repairs the down time and repair time could be weeks or even months. The consequences of some faults may be major contract works such as rewiring or large equipment replacement. In such cases careful planning and decisions must be taken by the contractor's management and design team when deciding to commit the company to carry out such works.

The following are basic considerations that any company should make before commencing such contracts:

- **Availability of staff** – Has the company enough people employed to carry out and complete the work?

- **Competency** – Does the company have staff with relevant experience?

- **Cost implications** – Has the company the monetary capability for the contract?

- **Special plant and machinery** – Has the company this type of resource available?

Legal responsibility

The legal responsibilities surrounding the replacement or repair of a fault must also be considered when making the decision to repair or replace. It is usual for a contract to be entered into by both parties as this would help to avoid, or at least assist, in settling future disputes. Such a contract would itemise:

- costs
- time
- guarantee period/warranty
- omissions from contract.

The costs should be agreed and adhered to. Most customers will not pay more than the initial agreed costs of repair unless an allowance was made in the contract for unforeseen work. Unforeseen work could result from something found when dismantling or taking out existing items, such as finding faulty or dangerous wiring underneath an item of heavy plant that is to be repaired.

If the initial costing accounted for repair and replacement of parts, then it is essential that you do what you said you were going to do. For instance, if an inspection is subsequently carried out and it is evident that no work has taken place then you may be liable for legal proceedings to be taken against you for a breach of contract terms.

An accurate cost analysis should be considered when giving an estimate or fixed price for repair work, as it is usual practice to carry out repair work on a daily rate where the customer pays only for the time taken and materials used in rectifying the fault.

The customer should also be given a guarantee for the work done, both in respect of quality and also for the materials used.

When deciding whether to repair or replace, it may be deemed necessary to avoid writing a guarantee in the contract when the client insists on re-using existing items in order to reduce costs. It is often the case that old industrial or antiquated items would be prone to failure and to avoid legal action such items should not be included in the contract and written out of your guarantee.

Other factors affecting the fault repair process

Careful planning can mean that most work is carried out efficiently with little or no disruption to the client. However, faults on installation equipment and wiring systems are certainly not planned and trained personnel can only minimise the disruption caused by the effects of faults.

In a factory environment, maintenance electricians are on hand to limit the effect of a fault and can usually solve problems and reduce the down time of a machine or

Remember

Repairs must not only be viable for the client but also for the contractor, who would not like to lose money or reputation on installation and repair works

Remember

Most companies will carry insurance to cover against any liabilities in respect of work done circuit to a minimum. This is because of their day-to-day working knowledge of the electrical services and the equipment connected to it.

When an electrician is called out to a fault they may not have the day-to-day knowledge or experience of the maintenance staff. However, through training and knowledge the electrician will use contingency plans to institute the procedures to rectify the fault with the minimum of disruption.

Such contingency plans provide a common format when dealing with fault finding at any customer's premises. Your company will have its own procedures and these may include:

- signing in
- wearing identification badges
- locating supervisory personnel
- locating data drawings etc.

- liaising with the client and office before commencing any work
- following safe isolation procedures.

Special requirements

We have previously looked at the safety procedures and the actions needed for determining causes of faults. If we assume that safety requirements and the location of faults have been met and that the prime objective of the electrical contractor (the electrician) is to locate the fault, rectify and re-commission, then this will be done in liaison with the customer and with minimal disruption. However the following topics are worthy of note.

Access to the system during normal working hours

There is nothing more costly or embarrassing than arriving at a customer's premises to carry out work of any kind, only to find the premises locked or unmanned because no prior arrangements were made for access. Access to the premises and the electrical installation or the specific item of equipment may prove a problem, especially if you are unfamiliar with the customer's premises.

When attending a fault or breakdown we must remember the procedures for ascertaining the information and familiarising ourselves with the installation. Remember our logical sequence was:

1. Identify the symptoms
2. Gather information
3. Analyse the evidence
4. Check supply
5. Check protective devices

6. Isolation and test
7. Interpret information and test results
8. Rectify the fault
9. Carry out functional tests
10. Restore the supply.

If we need to familiarise ourselves with the installation, then access must be readily available not only at the fault repair location but also at the supply intake and isolation point, and this has to be agreed prior to commencing any contracted work, whether it is a repair or a new installation.

Such prior agreement must take into account those personnel who may have their normal day-to-day routine and activities affected. It is not always possible to carry out repair work during normal working hours and so to avoid disruption it is often the case that such work is done outside of normal working hours. It is therefore important that arrangements for out-of-hours work are pre-arranged to account for security and especially access to the premises.

Is there a need for building fabric restoration?

As electricians, we are more often than not being asked to locate and repair faults on existing installations. Consequently, depending on the fault found and the method of rectification needed, we may need to disturb the fabric of the building. It is therefore vitally important to fully discuss with the client all of the implications of the fault diagnosis/rectification process. As part of investigating a fault, we may have to do nothing more than look inside an item of equipment. However, if the fault has been caused by someone putting a nail through a cable, then the repair will involve replacing the cable and that in turn may involve having to damage the building fabric. You should therefore explain these aspects to the client and agree what repair work (e.g. brickwork, plastering, decorating) will be required and which, if any, of these can be carried out satisfactorily by yourself.

Obviously, as with any aspect of fault finding and rectification, always leave the site in a clean and safe condition.

Whether the system can be isolated section by section

After access to the installation has been agreed, the repairs should be carried out using a logical approach. However, if the premises are still occupied and parts of the system are still being used, then it is important that the faulty circuit or plant to be worked on is isolated.

When only a small area or a single circuit or piece of equipment is affected it is not good practice to switch off the mains supply and isolate all the installation. By analysing the installation drawings and data the area or items should be isolated individually to limit disruption.

Provision of emergency or stand-by supply

Often the disruption to the customer is not having an electrical supply to an area of the installation. If the work cannot be done outside of normal working hours, a temporary supply could be arranged. Some installations may have a stand-by service in the form of emergency lighting or a stand-by generator, but this is only usual in hospitals or where computers are being used and data is stored.

Client demand for a continuous supply

Where the client does not want any interruption in supply and it is not possible to work on the affected area during normal working hours, arrangements should be made for a planned maintenance or shut down in order for the repair to be carried out safely.

Remember

The expertise of building contractors may be needed and there will be costs associated with this that need to be agreed in advance of any repairs being carried out

Remember

All electrical installation systems should have been designed and installed in accordance with the *IEE Wiring Regulations*, thus ensuring that the facility for isolating circuits and sections is inherent in the system

Remember

Any temporary supply which is provided should be wired in accordance with the *IEE Wiring Regulations*

Did you know?

Some formal risk assessments are recorded and noted in method statements. These can be seen in your company's Health and Safety Statements

Safety tip

The levels of light carried by fibre optic cabling can be dangerously high so you should never look into the end of a fibre optic cable

Special situations

Special precautions and a risk analysis should always be carried out before any work is commenced. It is a requirement of the Health and Safety at Work Act 1974 that precautions should be accounted for and that any risk is removed or persons carrying out any work activity should be protected from harm. Risk assessments are now commonplace and are part of all our daily work routine. In this section we shall be looking at the risks involved in carrying out fault diagnosis and repair.

Fibre optic cabling

This type of wiring system is becoming increasingly popular in the data and telecommunications industry, although it is unlikely that you will terminate or fault find on such systems. However, your company may be asked to install such systems and you may be asked to assist in their installation.

As its name suggests the cable is made up of high-quality minute glass fibres, which can carry vast amounts of transmitted signals and data using emitted signals of light. The levels of light transmitted can reach dangerously high levels. Therefore YOU SHOULD NEVER LOOK INTO THE END OF FIBRE OPTIC CABLE!

Antistatic precautions

Antistatic precautions are usually prevalent in areas that are hazardous due to the high risk of ignition or explosion. This is usually where flammable, ignitable liquids, gases or powders are stored or dispensed, for instance:

- petrol filling stations
- chemical works
- offshore installations
- paint stores
- flour mills.

In such situations we must be aware that NO NAKED FLAMES or equipment which can cause a spark or discharge of energy can be used. There would obviously be a no smoking policy. It is important to adhere to the safety signs and regulations in such areas. When asked to work in such environments you should be given special induction and training to make you aware of the dangers and the risks.

Electro-static discharge

Static is an electric charge, caused by an excess or deficiency of electrons collected on a conducting or non-conducting surface, thus creating a potential difference measured in volts. Static charges are caused in several ways, as described opposite.

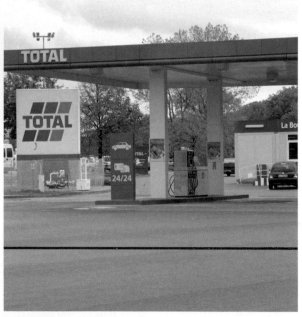

Petrol filling station

Friction

When two surfaces are rubbed together they behave as though electrons were rubbed off one surface and on to the other, leaving one surface with a deficiency of electrons, therefore being positively charged. The other surface acquires and retains electrons, therefore being negatively charged.

Separation

When self-adhesive plastic or cellulose tape is rapidly pulled off a reel it generates a static charge by tearing electrons from one surface to another.

Induction

Static is generated by induction when a charged item, for example a folder of paper, is placed on a desk and on top of that is placed a conductor, say a printed circuit board. If the paper was positively charged, the charge from the paper is immediately induced into the printed circuit board, making it negatively charged.

All materials can be involved in the production of static. In the case of non-conductors (insulators such as paper, plastic and textiles) the charge cannot redistribute itself and therefore remains on the surface.

Static electricity

Did you know?

The reality is that every step, even every shuffle on a seat, generates static charges. Picking up, putting down, wrapping and unwrapping, even just handling an object creates static. So you can see that with all the everyday movements we make, while wearing a lot of clothes made from man-made fibres, we will generate large amounts of static

Protection from electro-static discharge

The main effort must be to protect electro-static sensitive devices by providing special handling areas where they can be handled, stored, assembled, tested, packed and dispatched. In the special handling area there should be total removal of all untreated wrappings, paper, polystyrene, plastics, non-conductive bins, racks and trays, and soldering irons including tips should be earthed. Specifications, drawings and work instructions must be in protective bags, folders or bins. Wrist straps should be worn at all times in the static free area. These straps 'earth' the handler and provide a safe path for the removal of static electricity.

Damage to electronic devices due to over voltage

When testing or inspecting faults we must take account of electronic equipment and their rated voltage. It is usual to isolate or disconnect such equipment to avoid damage due to test voltages exceeding the equipment's rated voltage. Such voltage levels will cause components, control equipment, or data and telecommunication equipment to be faulty. This could be costly not only in repairs but also to the reputation of your company, and could result in the loss of future contracts. The voltage of an insulation resistance tester can reach levels between 250 to 1000 volts.

Avoidance of shut down of IT equipment

On no account should any circuit be isolated where computer equipment is connected. This type of equipment is to be found in many commercial, industrial and even domestic installations. The loss of data is the most common occurrence when systems fail due to faults or power supply problems. Larger organisations will have data protection in the form of uninterruptible power supplies (UPS systems). Whether UPS systems are in place or not, it is still important to plan the isolation of circuits where IT systems are connected. This will give the client time to arrange data storage and download, which will allow systems to be switched off and repairs to be carried out.

Risk of high frequency on high capacitive circuits

Capacitive circuits could be circuits with capacitors connected to them or some long runs of circuit wiring which may have a capacitive effect. This is usual on long runs of mineral-insulated cables. When working on such circuits no work should commence until the capacitive effect has been discharged. In some cases it would be practical to discharge capacitors manually by shorting out the capacitor.

Danger from storage batteries

Batteries can be found on many installations and provide an energy source in the event of an emergency when there is a power failure. This can be seen where battery panels provide a back-up for the emergency lighting in a building when the power fails.

Batteries can also be found in alarm panels, IT UPS systems and emergency stand-by generators for starting.

Wherever you have cause to work with or near batteries special care should be observed, particularly with regard to:

- Lead acid cells, which contain dilute sulphuric acid. This acid is harmful to the skin and eyes, can rot clothing and is highly corrosive. If there is any contact with the skin it should be immediately diluted and washed with water.

- Lead acid cells also emit hydrogen gas when charging and discharging, which when mixed with air is highly explosive.

- High voltages applied to cell terminals will damage the battery.

- When connecting cells, shorts across the terminals will produce arcing, which could cause an explosion.

- Cells should never be disconnected, as this can also produce arcing, which could cause an explosion.

Safety tip

On some systems where high frequencies exist, there is a danger of exceptionally high voltages occurring across the terminals. On no account should you as a trainee ever attempt to tamper with capacitive circuits and, if in doubt, seek the assistance of your supervisor

FAQ

Q   **Is there any other advice you can give me about fault finding ?**

A   Apart from using the information given in this chapter, the best advice that I ever received from an electrician was to stay calm and treat fault finding as being a process of elimination and simply 'looking for the differences between what you should see and what you are actually looking at'. Listen carefully to any fault being explained to you, it may help establish the cause more quickly. Wherever possible, look at the fault being demonstrated and if no circuit drawings exist, then make your own. You can then use the drawing to prove wiring and then tick them off on the drawing as you find them and prove their connections.

Q   **Why all the fuss about safe isolation if they said it was switched off ?**

A   Never take someone else's word that a circuit or piece of equipment is safe to work on. You should always make sure that, before you start work on any piece of equipment, it is completely isolated from the supply in a safe and approved manner. You really don't want the supply re-energised to something you are working on…do you?

Therefore, when you are visiting a client's premises, investigate the means of isolation thoroughly. It's no good going into a factory to repair a machine where every machine is controlled by a separate supply that is fully interlocked and padlocked if, when given the padlock key for that machine, you then find that every member of the maintenance team has a key to open every padlock. Equally, never trust circuit markings inside distribution boards. Just because you switched off the MCB that said it controlled the lighting, doesn't mean that it actually does.

Finally, always prove that a supply is not present with a suitable and functioning voltage indicator. Remember that the supply to a piece of equipment might pass through several other bits of equipment on its way to you and if any of them were switched off when you tested for a supply at the equipment, then you would not know whether you had isolated the correct circuit.

On the job: Fault diagnosis

You are asked by your employer to go to a factory unit to investigate a fault with a piece of equipment. When you arrive the customer takes you to the factory area and demonstrates the fault to you. The piece of equipment has a motor, which is fed from a distribution board via a DOL starter/isolator with a 230V coil, which is in turn linked to a remote stop/start control station located some 6 m away in a noise reduction area. The fault is that the motor will not start at all when controlled from the starter/isolator, but at the remote station when the start button is pressed, the motor starts but stops again as soon as you take your finger off the start button. The customer has a wiring diagram for the starter and remote control, so how would you try to solve this problem?

(A suggested answer can be found on page 315)

Knowledge check

1. What is the document published by the HSE that gives guidance for electrical test equipment used by electricians?

2. You are asked to repair a fluorescent lighting fitting that is fed from an MCB distribution board. Describe how you would safely isolate the circuit before starting work on it.

3. Give three means by which an installation designer can influence fault protection and damage limitation.

4. How would you define 'overload current', 'overcurrent', 'earth-fault current' and 'short circuit current'?

5. What is meant by 'transient voltages'?

6. How can regular inspection and maintenance assist in the prevention of faults?

7. A poor or loose connection can lead to what happening?

8. As an installation electrician you may be expected to find faults on a range of systems and equipment. However, what should you generally not attempt to diagnose faults on?

9. Knowledge of an installation is an essential requirement in fault diagnosis. Name four different people that you could approach to gain knowledge of an installation and a fault.

10. Describe the logical stages of fault diagnosis and rectification.

Chapter 7

Motors, generators and motor control

OVERVIEW

In essence there are two categories of motor: those that run on direct current (d.c.) and those that run on alternating current (a.c.).

To understand motors and generators we need to remember the basic principles of magnetic fields, current flow and induced motion and how when used together they can make a motor rotate.

These individual areas were covered in the Science section of Book 1, but we will look at them again in the context of their relationship to motors. This chapter will therefore cover the following:

- Concept revision
- The d.c. motor
- The d.c. and a.c. generator
- The a.c. motor
- Motor construction
- Motor speed and slip calculation
- Synchronous a.c. induction motors
- Motor windings
- Motor starters
- Motor speed control

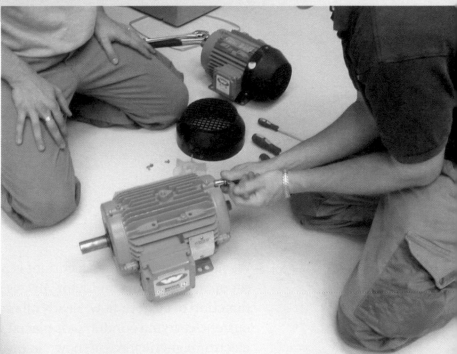

Concept revision

Magnetic fields

Between the poles of a simple permanent magnet, there is a magnetic field. The direction of the magnetic field moves from the north pole of the magnet to the south pole in closed loops. Like poles repel and unlike poles attract. The magnetic field is represented by lines of force – the magnetic flux (remember the iron filings experiment referred to in chapter 3 of Book 1).

Figure 7.1   Magnetic field around a bar magnet

Current flow and magnetic fields

Now, let's consider a conductor with an electric current flowing through it. When this happens a magnetic field will be produced in the form of concentric circles along the length of the conductor.

Figure 7.2   Magnetic field around a conductor

Force on a current-carrying conductor within a magnetic field

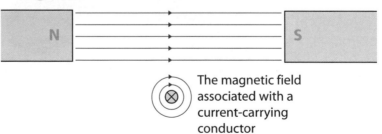

The magnetic field associated with a current-carrying conductor

Figure 7.3   The magnetic field associated with two fixed poles

We now know that when a current is present in a conductor a magnetic field is set up around that conductor, always in a clockwise direction, in relation to the direction of current flow. Nearly all motors work on the basic principle that when a current-carrying conductor is placed in a magnetic field it experiences a force. This electromagnetic force is shown in the following diagrams.

If we place the current-carrying conductor into the magnetic field as shown in Figure 7.3, you can see that the current is going away from you; therefore the field is clockwise. This can be best remembered by imagining a corkscrew being twisted into a cork; you need to turn the corkscrew to the right (this symbolises the magnetic field) and the corkscrew is moving away from you, which symbolises the current flowing away from you.

Direction of the force on the conductor

Figure 7.4 The magnetic field when the conductor is placed between the poles

In Figure 7.4 we should now note the following:

- The main field now becomes distorted.

- The field is weaker below the conductor because the two fields are in opposition.

- The field is stronger above the conductor because the two fields are in the same direction and aid each other. Consequently the force moves the conductor downwards.

In Figure 7.5 we can see that if either the current through the conductor or the direction of the magnetic field between the poles is reversed, the force acting on the conductor now tends to move it in the reverse direction.

Remember

⊗ indicates the current flowing away from you

⊙ indicates the current is flowing towards you

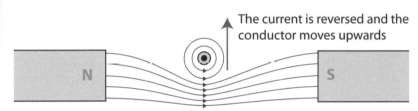

The current is reversed and the conductor moves upwards

Figure 7.5 Effect of reversing the direction of current or magnetic field

Fleming's left-hand rule

The direction in which a current-carrying conductor tends to move when it is placed in a magnetic field can be determined by **Fleming's left-hand (motor) rule**.

This rule states: if the first finger, the second finger and the thumb of the left hand are held at right angles to each other, as shown in Figure 7.6, then with the first finger pointing in the direction of the Field (N to S) and the second finger pointing in the direction of the current in the conductor, the thumb will now be indicating the direction in which the conductor tends to move.

First finger pointing in the direction of the field (N to S)

Second finger pointing in the direction of the current in the conductor

Thumb points in the direction in which the conductor tends to move

Figure 7.6 Fleming's left-hand (motor) rule

It is this basic principle of a force being exerted on a current-carrying conductor in a magnetic field that is used in the construction of motors and moving coil instruments.

So let's now look at a simple motor that operates using direct current (d.c.), in other words current flowing continuously in the same direction along a conductor, such as that produced by a battery.

Remember

The M in thuMb stands for Motion, the F in First finger stands for Field and the C in seCond finger stands for Current

The d.c. motor

Figure 7.7 Single current-carrying loop in a magnetic field

The operating principle of a d.c. motor is fairly straightforward. Looking at Figure 7.7, if we take a conductor, form it into a loop that is pivoted at its centre, place it within a magnetic field and then apply a d.c. supply to it, a current will pass around the loop. This will cause a magnetic field to be produced around the conductor, and this magnetic field will interact with the magnetic field between the two poles (**pole pair**) of the magnet.

The magnetic field between the two poles (pole pair) becomes distorted and, as mentioned earlier, this interaction between the two magnetic fields causes a bending or stretching of the lines of force. The lines of force behave a bit like an elastic band and consequently are always trying to find the shortest distance between the pole pair. These stretched lines, in their attempt to return to their shortest length, therefore exert a force on the conductor that tries to push it out of the magnetic field, and thus they cause the loop to rotate.

In reality we don't have just one loop, we have many, with them being wound on to a central rotating part of the d.c. motor known as the **armature**. When the armature is positioned so that the loop sides are at right angles to the magnetic field, a turning force is exerted. But we have a problem when the coil has rotated 180°, as the magnetic field in the loop is now opposite to that of the field, and this will tend to push the armature back the way it came, thus stopping the rotating motion.

The solution is to reverse the current in the armature every half rotation so that the magnetic fields will work together to maintain a continuous rotating motion. The device that we use to achieve this switching of polarity is known as the **commutator**.

The commutator

Figure 7.8 Single loop with commutator

Figure 7.8 now gives a different perspective of the same arrangement, showing the device that enables the supply to be connected to the loop, enabling the loop to rotate continuously. This is the commutator. This simplified version has only two segments, connecting to either side of the loop, but in actual machines there may be any number. Large machines would have in excess of 50, and the numbers of loops can be in the hundreds. The commutator is made of copper, with the segments separated from each other by insulation.

Figure 7.9 now shows the direction of current around the loop when connected to a d.c. supply. The brushes remain in a fixed position, against the copper segments of the commutator. Note the direction of current in those sections indicated by X and Y in the diagram.

Figure 7.10 now shows the wire loop having rotated 180°. X and Y have changed positions but, as can be seen from the arrows, current flow remains as it was for the previous diagram. As the poles of the armature electromagnet pass the poles of the permanent magnets, the commutator reverses the polarity of the armature electromagnet. In that instant of switching polarity, inertia keeps the motor going in the proper direction and thus the motor continues to rotate in one direction.

In the previous drawings, we have shown the armature rotating between a pair of magnetic poles. Practical d.c. motors do not use permanent magnets, but use electromagnets instead. The electromagnet has two advantages over the permanent magnet:

1. By adjusting the amount of current flowing through the wire the strength of the electromagnet can be controlled.

2. By changing the direction of current flow the poles of the electromagnet can be reversed.

Figure 7.9 Single loop with d.c. flowing

Reversing a d.c. motor

The direction of rotation of a d.c. motor may be reversed by either:

- reversing the direction of the current through the field hence changing the field polarity.

- reversing the direction of the current through the armature.

Common practice is to reverse the current through the armature, and this is normally achieved by reversing the armature connections only.

Types of d.c. motor

Figure 7.10 Single loop rotated 180°

There are three basic forms of d.c. motor:

- series
- shunt
- compound.

They are very similar to look at, the difference being the way in which the field coil and armature coil circuits are wired.

The series motor has the field coil wired in series with the armature. It is also called a universal motor because it can be used in both d.c. and a.c. situations, has a high starting torque (rotational force) and a variable speed characteristic. The motor can therefore start heavy loads, but the speed will increase as the load is decreased.

The shunt motor has the armature and field circuits wired in parallel, and this gives constant field strength and motor speed.

The compound motor combines the characteristics of both the series and the shunt motors and thus has high starting torque and fairly good speed torque characteristics. However, because it is complex to control, this arrangement is usually only used on large bi-directional motors.

The d.c. and a.c. generator

We now know that when a current is present in a conductor a magnetic field is set up around that conductor, always in a clockwise direction, in relation to the direction of current flow.

We also know that when a conductor passes at right angles through a magnetic field, current is induced into the conductor; the direction of the induced current will depend on the direction of movement of the conductor, and the strength of the current will be determined by the speed at which the conductor moves.

It therefore follows that if we were to take this arrangement and connect it to some device that would spin the wire loop within the permanent magnetic field, we would then induce into the wire loop an e.m.f., and were we to connect a load to the armature via the commutator and brushes, a current would flow around the circuit and the load would work. In other words, we would have created a generator.

A variety of sources can be used mechanically to turn the generator's armature, such as steam, wind, water fall or petrol/diesel driven motors. Figure 7.11 below shows just such an arrangement and the voltage output for one complete revolution.

Figure 7.11 Voltage output for one complete revolution

As you can see from Figure 7.11, the output from such a generator has no negative parts in its cycle. This type of generator produces a voltage/current that alternates in magnitude but flows in one direction only: in other words we have a direct current (d.c.).

When a d.c. generator contains only a single coil it provides a pulsating d.c. output, as shown by the wave form in Figure 7.11. Consequently, in general use a number of coils are used to produce a more stable output.

The operating principle of the a.c. generator is much the same as that of the d.c. version. However, instead of the loop ends terminating at the commutator, they are terminated at slip rings. The a.c. generator normally has a stationary armature known as the **stator** and a rotating magnetic field, known as the **rotor**.

The a.c. motor

A simple a.c. motor

As we mentioned earlier, there are three general types of d.c. motor. However, there are many types of a.c. motor, each one having a specific set of operating characteristics such as **torque**, speed, single-phase or three-phase, and this determines their selection for use. We can essentially group them in to two categories: single-phase and three-phase.

In a d.c. motor, electrical power is conducted directly to the armature through brushes and a commutator. Due to the nature of an alternating current, an a.c. motor doesn't need a commutator to reverse the polarity of the current. Whereas a d.c. motor works by changing the polarity of the current running through the armature (the rotating part of the motor), the a.c. motor works by changing the polarity of the current running through the stator (the stationary part of the motor).

The series-wound (universal) motor

We will begin the discussion of a.c. motors by looking at the series-wound (universal) motor because it is different in its construction and operation from the other a.c. motors considered here, and also because it is constructed as we have previously discussed in the Types of d.c. motor section, having field windings, brushes, commutator and an armature.

As can be seen in Figure 7.12, because of its series connection, current passing through the field windings also passes through the armature. The turning motion (torque) is produced as a result of the interaction between the magnetic field in the field windings and the magnetic field produced in the armature.

Field windings made from fine copper conductors

Commutator connected to 4 armature windings

Carbon brushes

Commutator and armature mounted on carbon steel shaft

a.c. supply

Laminated pole pieces

Figure 7.12 Series universal motor

For this motor to be able to run on an a.c. supply, modifications are made both to the field windings and armature **formers**. These are heavily laminated to reduce eddy currents and I^2R losses, which reduces the heat generated by the normal working of the motor, thus making the motor more efficient.

This type of motor is generally small (less than a kilowatt) and is used to drive small hand tools such as drills, vacuum cleaners and washing machines.

A disadvantage of this motor is that it relies on contact with the armature via a system of carbon brushes and a commutator. It is this point that is the machine's weakness, as much heat is generated through the arcing that appears across the gap between the brushes and the commutator. The brushes are spring-loaded to keep this gap to a minimum, but even so the heat and friction eventually cause the brushes to wear down and the gap to increase. These then need to be replaced, otherwise the heat generated as the gap gets larger will eventually cause the motor to fail.

The advantages of this machine are:

- more power for a given size than any other normal a.c. motor
- high starting torque
- relative cheapness to produce.

Three-phase induction motors

Three-phase induction motor showing component parts

Induction motors operate because a moving magnetic field induces a current to flow in the rotor. This current in the rotor then creates a second magnetic field, which combines with the field from the stator windings to exert a force on the rotor conductors, thus turning the rotor.

Production of the rotating field

Figure 7.13 shows the stator of a three-phase motor to which a three-phase supply is connected. The windings in the diagram are in star formation and two windings of each phase are wound in the same direction.

Figure 7.13 Stator of three-phase motor with three-phase supply connection

Each pair of windings will produce a magnetic field, the strength of which will depend upon the current in that particular phase at any instant of time. When the current is zero, the magnetic field will be zero. Maximum current will produce the maximum magnetic field.

As the currents in the three phases are 120° out of phase (see graph in Figure 7.13) the magnetic fields produced will also be 120° out of phase. The magnetic field set up by the three phase currents will therefore give the appearance of rotating clockwise around the stator.

The resultant magnetic field produced by the three phases is at any instant of time in the direction shown by the arrow in the diagram, where diagrams (1) to (7) in Figure 7.13 show how the direction of the magnetic field changes at intervals of 60° through one complete cycle. The speed of rotation of the magnetic field depends upon the supply frequency and the number of 'pole pairs', and is referred to as the **synchronous** speed. We will discuss synchronous speed later in this chapter.

The direction in which the magnetic field rotates is dependent upon the sequence in which the phases are connected to the windings. Reversing the connection of any two incoming phases can therefore reverse rotation of the magnetic field.

Stator construction

Figure 7.14 Stator construction

Stator field winding

As shown in Figure 7.14, the stator (stationary component) comprises the field windings, which are many turns of very fine copper wire wound on to formers, which are then fixed to the inside of the stator steel frame (sometimes called the yoke).

The formers have two roles:

1. to contain the conductors of the winding

2. to concentrate the magnetic lines of flux to improve the flux linkage.

The formers are made of laminated silicon steel sections to reduce eddy currents, thereby reducing the I^2R losses and reducing heat. The number of poles fitted will determine the speed of the motor.

Rotor construction and principle of operation

Essentially there are two main types of rotor:

1. squirrel-cage rotor

2. wound rotor.

Squirrel-cage rotor

In the squirrel cage (see Figure 7.15), the bars of the rotor are shorted out at each end by 'end rings' to form the shape of a cage. This shape creates numerous circuits within it for the induced e.m.f. and resultant current to flow and thus produce the required magnetic field.

Tinned bars shorted out at each end by a tinned copper end ring

Figure 7.15 Squirrel-cage rotor

Figure 7.16 shows the cage fitted to the shaft of the motor. The rotor bars are encased within many hundreds of very thin laminated (insulated) segments of silicon steel and are skewed to increase the rotor resistance.

Carbon steel shaft

Tinned copper rotor bars encased in thin 0.5 mm laminated steel segments, to reduce losses and skewed to assist starting

Low-friction bearings

Tinned copper end rings short out the rotor bars

Figure 7.16   Cage fitted to shaft and motor

On the shaft will be two low-friction bearings which enable the rotor to spin freely. The bearings and the rotor will be held in place within the yoke of the stator by two end caps which are normally secured in place by long nuts and bolts that pass completely through the stator.

Figure 7.17   Rotating field

When a three-phase supply is connected to the field windings, the lines of magnetic flux produced in the stator, rotating at 50 revolutions per second, cut through the bars of the rotor, inducing an e.m.f. into the bars.

Faraday's Law states that 'when a conductor cuts, or is cut, by a magnetic field, then an e.m.f. is induced in that conductor the magnitude of which is proportional to the rate at which the conductor cuts or is cut by the magnetic flux.'

Did you know?

Squirrel cage induction motors are sometimes referred to as the 'workhorse' of the industry, as they are inexpensive and reliable and suited to most applications

The e.m.f. produces circulatory currents within the rotor bars, which in turn result in the production of a magnetic field around the bars. This leads to the distortion of the magnetic field as shown in Figure 7.17. The interaction of these two magnetic fields results in a force being applied to the rotor bars, and the rotor begins to turn. This turning force is known as a torque, the direction of which is always as Fleming's left-hand rule indicates.

Wound rotor

In the wound rotor type of motor, the rotor conductors form a three-phase winding, which is starred internally. The other three ends of the windings are brought out to slip rings mounted on the shaft. Thus it is possible through brush connections to introduce resistance into the rotor circuit, albeit this is normally done on starting only to increase the starting torque. This type of motor is commonly referred to as a slip-ring motor.

Figure 7.18 shows a completed wound-rotor motor assembly. Although it looks like a squirrel-cage motor, the difference is that the rotor bars are exchanged for heavy conductors that run through the laminated steel rotor, the ends then being brought out through the shaft to the slip rings on the end.

Figure 7.18 Wound-rotor motor assembly

The wound-rotor motor is particularly effective in applications where using a squirrel-cage motor may result in a starting current that's too high for the capacity of the power system. The wound-rotor motor is also appropriate for high-inertia loads having a long acceleration time. This is because you can control the speed, torque and resulting heating of the wound-rotor motor. This control can be automatic or manual. It's also effective with high-slip loads as well as adjustable-speed installations that do not require precise speed control or regulation. Typical applications include conveyor belts, hoists and elevators.

Single-phase induction motors

Figure 7.19 Single-phase induction motor

If we were to construct an induction motor as shown in Figure 7.19, we would find that, on connecting a supply to it, it would not run. However, if we were then to spin the shaft with our fingers, we would find that the motor *would* continue to run. Why is this?

When an a.c. supply is connected to the motor, the resulting current flow, and therefore the magnetic fields produced in the field windings, changes polarity, backwards and forwards, 100 times per second. Therefore no lines of flux cut through the rotor bars. If no lines of flux cut through the rotor bars then there is no e.m.f. being produced in them and there is therefore no magnetic field for the stator winding to interact with.

However, if we spin the rotor, we create the effect of the rotor bars cutting through the lines of force, hence the process starts and the motor runs up to speed. If we stop the motor and then connect the supply again, the motor will still not run. If this time we spin the rotor in the opposite direction, the motor will run up to speed in the new direction. So how can we get the rotor to turn on its own?

If we think back to the three-phase motor we discussed previously, we did not have this problem, because the connection of a three-phase supply to the stator automatically produced a rotating magnetic field. This is what is missing from the motor in Figure 7.19: we have no rotating field.

The split-phase motor (induction start/induction run)

We can overcome this problem if we add another set of poles, positioned 90° around the stator from our original wiring, as shown in Figure 7.20.

Now when the supply is connected, both sets of windings are energised, both windings having resistive and reactive components to them – resistance as every conductor has and also inductive reactance because the conductors form a coil. (These concepts were discussed more fully in Chapter 1 Electrical science.) These are known as the 'start' and 'run' windings.

The start winding is wound with fewer turns of smaller wire than the main winding, so it has a higher resistance. This extra resistance creates a small phase shift that results in the current in the start winding lagging that in the run winding by approximately 30°, as shown in Figure 7.21.

Start-winding high-resistance smaller csa conductors compared with the run winding

Run-winding low-resistance copper conductors

a.c. supply

Figure 7.20 Split-phase induction motor

Strong magnetic field in the run winding, weaker field in the start winding

Strong magnetic field in the start winding 30° later, weaker field in the run winding

Figure 7.21 Run and start winding phases

Therefore the magnetic flux in each of the windings is growing and collapsing at different periods in time, so that for example as the run winding is having a strong north/south on the face of its pole pieces, the start winding will only have a weak magnetic field.

In the next instance, the run winding's magnetic field has started to fade, but the start winding's magnetic field is now strong, presenting to the rotor an apparent shift in the lines of magnetic flux.

In the next half of the supply cycle, the polarity is reversed and the process repeated, so there now appears to be a rotating magnetic field. The lines of force cutting through the rotor bars induce an e.m.f. into them, and the resulting current flow now produces a magnetic field around the rotor bars: the interaction between the magnetic fields of the rotor and stator takes place and the motor begins to turn.

It is because the start and run windings carry currents that are out of phase with each other that this type of motor is called the 'split-phase'.

Once the motor is rotating at about 75 per cent of its full load speed, the start winding is disconnected by the use of a device called a centrifugal switch, which is attached to the shaft; see Figure 7.22.

Centrifugal contacts in start winding

Start-winding high-resistance smaller csa conductors compared with the run winding

Run-winding low-resistance copper conductors

a.c. supply

Figure 7.22   Split-phase induction motor with centrifugal switch

This switch works by centrifugal action, in that sets of contacts are held closed by a spring, and this completes the circuit to the start winding. When the motor starts to turn, a little weight gets progressively thrown away from the shaft, forcing the contacts to open and thus disconnecting the start winding. It's a bit like the fairground ride known as 'the Rotor', in which you are eventually held against the sides of the ride by the increasing speed of the spinning wheel. Once the machine has disconnected the start winding, the machine continues to operate from the run winding.

The split-phase motor's simple design makes it typically less expensive than other single-phase motors. However, it also limits performance. Starting torque is low at about 150 per cent to 175 per cent of the rated load. Also, the motor develops high starting currents of about six to nine times the full load current. A lengthy starting time can cause the start winding to overheat and fail, and therefore this type of motor shouldn't be used when a high starting torque is needed. Consequently it is used on light-load applications such as small hand tools, small grinders and fans, where there are frequent stop/starts and the full load is applied after the motor has reached its operating speed.

Reversal of direction

If you think back to the start of this section, we talked about starting the motor by spinning the shaft. We also said that we were able to spin it in either direction and the motor would run in that direction. It therefore seems logical that in order to change the direction of the motor, all we have to concern ourselves with is the start winding. We therefore need only to reverse the connections to the start winding to change its polarity, although you may choose to reverse the polarity of the run winding instead. The important thing to remember is that if you change the polarity through both the run and start windings, the motor will continue to revolve in the same direction.

The capacitor-start motor (capacitor start/induction run)

Normally perceived as being a wide-ranging industrial motor, the capacitor-start motor is very similar to the split-phase motor discussed previously. Indeed, it probably helps to think of this motor as being a split-phase motor but with an enhanced start winding that includes a capacitor in the circuit to help out with the start process. If we look at the photograph we can see the capacitor mounted on top of the motor case.

In this motor the start winding has a capacitor connected in series with it, and since this gives a phase difference of nearly 90° between the two currents in the windings, the starting performance is improved. We can see this represented in the sine waves shown in Figure 7.23.

Capacitor-start motor

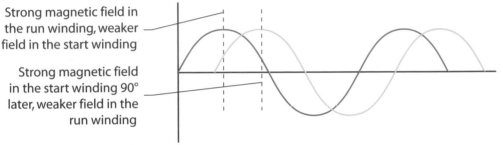

Strong magnetic field in the run winding, weaker field in the start winding

Strong magnetic field in the start winding 90° later, weaker field in the run winding

Figure 7.23   Magnetic field in capacitor-start motor

In this motor the current through the run winding lags the supply voltage due to the high inductive reactance of this winding, and the current through the start winding leads the supply voltage due to the capacitive reactance of the capacitor. The phase displacement in the currents of the two windings is now approximately 90°. Figure 7.24 shows the winding connections for a capacitor-start motor.

Figure 7.24   Winding connections for capacitor-start split-phase motor

The magnetic flux set up by the two windings is much greater at starting than in the standard split-phase motor, and this produces a greater starting torque. The typical starting torque for this type of motor is about 300 per cent of full-load torque, and a typical starting current is about five to nine times the full-load current.

The capacitor-start motor is more expensive than a comparable split-phase design because of the additional cost of the capacitor. But the application range is much wider because of higher starting torque and lower starting current. Therefore, because of its improved starting ability, this type of motor is recommended for loads that are hard to start, so we see this type of motor used to drive equipment such as lathes, compressors and small conveyor systems.

As with the standard split-phase motor, the start windings and the capacitor are disconnected from the circuit by an automatic centrifugal switch when the motor reaches about 75 per cent of its rated full-load speed.

Reversal of direction

Reversing the connections to the start winding will only change its polarity, although we may choose to reverse the polarity of the run winding instead.

Permanent split capacitor (PSC) motors

Permanent split capacitor (PSC) motors look exactly the same as capacitor-start motors. However, a PSC motor doesn't have either a starting switch or a capacitor that is strictly used for starting. Instead, it has a run-type capacitor permanently connected in series with the start winding, and the second winding is permanently connected to the power source. This makes the start winding an auxiliary winding once the motor reaches running speed. However, because the run capacitor must be designed for continuous use, it cannot provide the starting boost of a starting capacitor.

Typical starting torques for thus type of motor are low, from 30 to 150 per cent of rated load, so these motors are not used in difficult starting applications. However, unlike the split-phase motor, PSC motors have low starting currents, usually less than 200 per cent of rated load current, making them excellent for applications with high cycle rates.

PSC motors have several advantages. They need no starting mechanism and so can be reversed easily, and designs can easily be altered for use with speed controllers. They can also be designed for optimum efficiency and high power factor at rated load.

Permanent split capacitor motors have a wide variety of applications depending on the design. These include fans, blowers with low starting torque and intermittent cycling uses such as adjusting mechanisms, gate operators and garage-door openers, many of which also need instant reversing.

Capacitor start–capacitor run motors

Capacitor start–capacitor run motor

In appearance we can distinguish this motor because of the two capacitors that are mounted on the motor case.

This type of motor is widely held to be the most efficient single-phase induction motor, as it combines the best of the capacitor-start and the permanent split capacitor designs and is able to handle applications too demanding for any other kind of single-phase motor.

As shown in Figure 7.25, it has a start capacitor in series with the auxiliary winding like the capacitor-start motor, and this allows for high starting torque. However, like the PSC motor, it also has a run capacitor that remains in series with the auxiliary winding after the start capacitor is switched out of the circuit.

Another advantage of this type of motor is that it can be designed for lower full-load currents and higher efficiency, which means that it operates at a lower temperature than other single-phase motor types of comparable horsepower. Typical uses include woodworking machinery, air compressors, high-pressure water pumps, vacuum pumps and other high-torque applications.

Figure 7.25 Capacitor start–capacitor run split-phase motor

Shaded pole motors

One final type of single-phase induction motor that is worthy of mention is the shaded-pole type. We cover this last as, unlike all of the previous single-phase motors we have discussed, shaded-pole motors have only one main winding and no start winding.

Starting is by means of a continuous copper loop wound around a small section of each motor pole. This 'shades' that portion of the pole, causing the magnetic field in the ringed area to lag the field in the non-ringed section. The reaction of the two fields then starts the shaft rotating.

Because the shaded pole motor lacks a start winding, starting switch or capacitor, it is electrically simple and inexpensive. In addition, speed can be controlled merely by varying voltage or through a multi-tap winding.

The shaded pole motor has many positive features, but it also has several disadvantages. As the phase displacement is very small, it has a low starting torque, typically in the region of 25 to 75 per cent of full-load torque. Also, it is very inefficient, usually below 20 per cent.

Low initial costs make shaded pole motors suitable for light-duty applications such as multi-speed fans for household use and record turntables.

Motor construction

To help you understand how motors are put together, this section includes photographs and diagrams of the components used in motor construction.

The three-phase squirrel-cage induction motor

Components of a three-phase squirrel-cage induction motor

Field winding

Figure 7.26 Construction of a three-phase squirrel-cage induction motor

The single-phase induction motor

This motor consists of a laminated stator wound with single-phase windings arranged for split-phase starting, and a cage rotor. The cage rotor is a laminated framework with lightly insulated longitudinal slots into which copper or aluminium bars are fitted.

The bars are then connected together at their ends by metal end-rings. No electrical connections are made to the rotor.

Figure 7.27 Construction of a single-phase induction motor

Motor speed and slip calculation

Speed of a motor

There are essentially two ways to express the speed of a motor.

- **Synchronous speed**. For an a.c. motor this is the speed of rotation of the stator's magnetic field. Consequently, this is really only a theoretical speed, as the rotor will always turn at a slightly slower rate.

- **Actual speed**. This is the speed at which the shaft rotates. The nameplate on most a.c. motors will give the actual motor speed rather than the synchronous speed.

The difference between the speed of the rotor and the synchronous speed of the rotating magnetic field is known as the **slip**, which can be expressed either as a unit or in percentage terms. Because of this, we often refer to the induction motor as being an asynchronous motor.

Remember, the speed of the rotating magnetic field is known as the synchronous speed, and this will be determined by the frequency of the supply and the number of pairs of poles within the machine. The speed at which the rotor turns will be between two per cent and five per cent slower, with an average of four per cent being common.

The reduced speed of the rotor is due to it having to overcome friction, as during the turning movement there is friction from the bearings in addition to any friction deriving from the load that the motor is connected to. Another factor that comes into play in determining the speed of a motor is **windage**. This means that within the enclosure there is a certain amount of air, and as the rotor rotates it has to move the air, which in turn contributes to the slowing down of the rotor. Of course there are no moving parts involved with the rotating magnetic field, so the rotor will never catch up. However, were some miracle to happen and the rotor reached the synchronous speed, we would have a different problem.

We already know that when a conductor passes at right angles through a magnetic field current is induced into the conductor. The direction of the induced current is dependant on the direction of movement of the conductor, and the strength of the current is determined by the speed at which the conductor moves. If the rotating magnetic field and the rotor are now revolving at the same speed, there will be no lines of magnetic flux cutting through the rotor bars, no induced e.m.f. and consequently no resultant magnetic field around the rotor bars to interact with the rotating magnetic field of the stator. The motor will immediately slow down and, having slowed down, will then start to speed up as the lines of magnetic flux start to cut through the rotor bars again – and so the process would continue.

Standard a.c. induction motors therefore depend on the rotor trying, but never quite managing, to catch up with the stator's magnetic field. The rotor speed is just slow enough to cause the proper amount of rotor current to flow, so that the resulting torque is sufficient to overcome windage and friction losses and drive the load.

Definition

Windage – the air resistance of a moving object or the force of the wind on a stationary object

Determining synchronous speed and slip

All a.c. motors are designed with various numbers of magnetic poles. Standard motors have two, four, six or eight poles, and these poles play an important role in determining the synchronous speed of an a.c. motor.

As we said before, the synchronous speed can be determined by the frequency of the supply and the number of pairs of poles within the machine. We can express this relationship with the following formula:

$$\text{Synchronous speed } (n_s) \text{ in revolutions per second} = \frac{\text{Frequency (f) in Hz}}{\text{The number of pole pairs (p)}}$$

Example 1

Calculate the synchronous speed of a four-pole machine connected to a 50 Hz supply.

$$n_s = \frac{f}{P}$$

As we know the motor has four poles, this means it has two pole pairs. We can therefore complete the calculation as:

$$n_s = \frac{50}{2} \qquad \text{therefore } n_s = 25 \text{ revolutions per second (rps)}$$

To convert revolutions per second into the more commonly used revolutions per minute (rpm), simply multiply n_s by 60. This new value is referred to as N_s, which in this example will become 25×60, giving 1500 rpm.

We also said that we refer to the difference between the speed of the rotor and the synchronous speed of the rotating magnetic field as the slip, which can be expressed either as a unit (S) or in percentage terms (S per cent). We express this relationship with the following formula:

$$\text{per cent slip} = \frac{\text{Synchronous speed } (n_s) - \text{Rotor speed } (n_r)}{\text{Synchronous speed } (n_s)} \times 100$$

Example 2

In this example numbers have been rounded up for ease.

A six-pole cage-induction motor runs at 4 per cent slip.

Calculate the motor speed if the supply frequency is 50 Hz.

$$S \text{ (per cent)} = \frac{(n_s - n_r)}{n_s} \times 100$$

We therefore need first to establish the synchronous speed; as the motor has six poles it will have three pole pairs. Consequently:

Synchronous speed $n_s = \dfrac{f}{P}$ giving us $\dfrac{50}{3}$ and therefore $n_s = 16.7$ revs/sec

We can now put this value into our formula and then, by transposition, rearrange the formula to make n_r the subject. Consequently:

$$S \text{ (per cent)} = \frac{(n_s - n_r)}{n_s} \times 100 \quad \text{giving us} \quad 4 = \frac{(16.7 - n_r)}{16.7} \times 100$$

Therefore by transposition:

$$4 = \frac{(16.7 - n_r)}{16.7} \times 100 \qquad \text{gives us} \qquad (16.7 - n_r) = \frac{4 \times 16.7}{100}$$

When calculated :

$$(16.7 - n_r) = \frac{4 \times 16.7}{100} \qquad \text{becomes} \qquad (16.7 - n_r) = 0.668$$

Therefore by further transposition:

$$(16.7 - n_r) = 0.668 \qquad \text{becomes} \qquad 16.7 - 0.668 = n_r$$

Therefore $n_r = \textbf{16.032 rps}$ or $N_r = \textbf{962 rpm}$

Synchronous a.c. induction motors

A synchronous motor, as the name suggests, runs at synchronous speed. Because of the problems discussed earlier, this type of motor is not self-starting and instead must be brought up to almost synchronous speed by some other means.

Three-phase a.c. synchronous motors

To understand how the synchronous motor works, assume that we have supplied three-phase a.c. power to the stator, which in turn causes a rotating magnetic field to be set up around the rotor. The rotor is then supplied via a field winding with d.c. and consequently acts a bit like a bar magnet, having north and south poles. The rotating magnetic field now attracts the rotor field that was activated by the d.c. This results in a strong turning force on the rotor shaft, and the rotor is therefore able to turn a load as it rotates in step with the rotating magnetic field.

It works this way once it's started. However, one of the disadvantages of a synchronous motor is that it cannot be started from a standstill by just applying a three-phase a.c. supply to the stator. When a.c. is applied to the stator, a high-speed rotating magnetic field appears immediately. This rotating field rushes past the rotor poles so quickly that the rotor does not have a chance to get started. In effect, the rotor is repelled first in one direction and then the other.

An induction winding (squirrel-cage type) is therefore added to the rotor of a synchronous motor to cause it to start, effectively meaning that the motor is started as an induction motor. Once the motor reaches synchronous speed, no current is induced in the squirrel-cage winding, so it has little effect on the synchronous operation of the motor.

Synchronous motors are commonly driven by transistorised variable-frequency drives.

Single-phase a.c. synchronous motors

Small single-phase a.c. motors can be designed with magnetised rotors. The rotors in these motors do not require any induced current so they do not slip backwards against the mains frequency. Instead, they rotate synchronously with the mains frequency. Because of their highly accurate speed, such motors are usually used to power mechanical clocks, audio turntables and tape drives; formerly they were also widely used in accurate timing instruments such as strip-chart recorders or telescope drive mechanisms. The shaded-pole synchronous motor is one version.

As with the three-phase version, inertia makes it difficult to accelerate the rotor instantly from stopped to synchronous speed, and the motors normally require some sort of special feature to get started. Various designs use a small induction motor (which may share the same field coils and rotor as the synchronous motor) or a very light rotor with a one-way mechanism (to ensure that the rotor starts in the 'forward' direction).

Motor windings

A motor can be manufactured with the windings internally connected. If this is the case and there are three terminal connections in the terminal block labelled U, V and W, you would expect the motor windings to be connected in a delta configuration. This is shown in Figure 7.28.

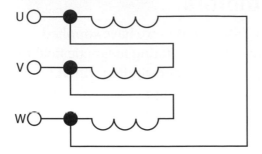

Figure 7.28   Motor windings with delta connection

However, there may be four connections in the terminal box labelled U, V, W and N. If this is the case, the windings would be arranged to give a star configuration, as shown in Figure 7.29.

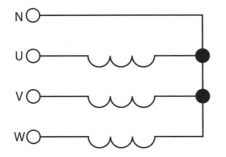

Figure 7.29   Motor windings with four connections

Alternatively the terminal block may contain six connections: U1, U2, V1, V2, W1 and W2. This is used where both star and delta configurations are required. The terminal connections can then be reconfigured for either star or delta, starting within the terminal block. Figure 7.30 illustrates the connections that would come out to the terminal block.

Figure 7.30   Motor windings with six connections

Motor starters

A practical motor starter has to do more than just switch on the supply. It also has to have provision for automatically disconnecting the motor in the event of overloads or other faults. The overload protective device monitors the current consumption of the motor and is set to a predetermined value that the motor can safely handle. When a condition occurs that exceeds the set value, the overload device opens the motor starter control circuit and the motor is turned off. The overload protection can come in a variety of types, including solid-state electronic devices.

The starter should also prevent automatic restarting should the motor stop because of supply failure. This is called **no-volts protection** and will be discussed later in this section.

The starter must also provide for the efficient stopping of the motor by the user. Provision for this is made by the connection of remote stop buttons and safety interlock switches where required.

The Direct-On-Line (DOL) starter

This is the simplest and cheapest method of starting squirrel-cage (induction) motors.

The expression 'Direct-On-Line' starting means that the full supply voltage is directly connected to the stator of the motor by means of a contactor-starter, as shown in Figure 7.31.

DOL starter

Figure 7.31   DOL starter

Since the motor is at a standstill when the supply is first switched on, the initial starting current is heavy. This 'inrush' of current can be as high as 6 to 10 times the full load current, i.e. a motor rated at 10A could have a starting current inrush as high as 60A, and the initial starting torque can be about 150 per cent of the full-load torque. Thus you may observe motors 'jumping' on starting if their mountings are not secure. As a result, Direct-On-Line starting is usually restricted to comparatively small motors with outputs of up to about 5 kW.

DOL starters should also incorporate a means of overload protection, which can be operated by either magnetic or thermal overload trips. These activate when there is a sustained increase in current flow.

To reverse the motor you need to interchange any two of the incoming three-phase supply leads. If a further two leads are interchanged then the motor will rotate in the original direction.

Operating principle of a DOL starter

A three-pole switch controls the three-phase supply to the starter. This switch normally includes fuses, which provide a means of electrical isolation and also short-circuit protection. We shall look at the operation of the DOL starter in stages.

Let's start by looking at the one-way switch again. In this circuit (Figure 7.32), the switch is operated by your finger and the contacts are then held in place mechanically.

We could decide that we don't want to operate the switch this way and instead use a relay. In this system (Figure 7.33), when the coil is energised it creates a magnetic field. Everything in the magnetic field will be pulled in the direction of the arrow and the metal strip will be pulled onto the contacts. As long as the coil remains energised and is a magnet, the light will stay on.

Figure 7.32   One-way switch

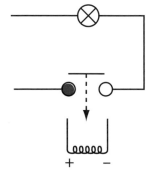

Figure 7.33   Switch relay

Looking at Figures 7.32 and 7.33, we can see that the first option works well enough for its intended purpose. However, it couldn't be used in a three-phase system as we would need one for each phase and would have to trust to luck each time we tried to hit all three switches at the same time. However, the second option does provide us with an effective method of controlling more than one thing from one switch, as long as they are all in the same magnetic field.

Let's apply this knowledge to a DOL starter. We know that we can't have three one-way switches in the starter. But it helps to try and think of the contacts: where the switches aren't operated by your fingers, but are pulled in the direction of the arrow by the magnetic effect of the coil (see Figure 7.34), items affected by the same magnetic effect are normally shown linked by a dotted line. For ease of explanation we'll use a 230V coil.

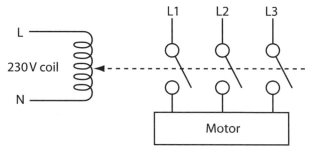

Figure 7.34   Three-switch relay

From this we can see that as soon as we put a supply onto the coil it energises, becomes a magnet and pulls the contacts in. Obviously, this would be no good for our starter, as every time the power is put on, the starter will become active and start operating whatever is connected to it. In the case of machinery this could be very dangerous. The starter design therefore goes one step further to include no-volts protection. The simple addition of a 'normally open' Start button gives this facility, as shown in Figure 7.35.

Figure 7.35 Three-switch relay with 'normally open' Start button

So far so good. We now have to make a conscious decision to start the motor.

Our next problem though, is that every time we take our finger off the Start button, the button springs back out, the supply to the coil is lost and the motor stops. This is the 'no volt' protection element in operation.

What we need for normal operation is a device called the 'hold in' or 'retaining' contact, as shown in Figure 7.36. This is a 'normally closed' contact (position 1) which is also placed in the magnetic field of the coil. Consequently, when the coil is energised, it is also pulled in the direction shown, and in this case, across and onto the terminals of the Start button (position 2).

We can now take our finger off the Start button, as the supply will continue to feed the coil by running through the 'hold in' contact that is linking out the Start button terminals (position 2).

This now means that we can only break the coil supply and stop the motor by fitting a Stop button. In the case of starters fitted with a 230 V coil, this will be a normally closed button placed in the neutral wire.

Now, for the fraction of a second that we push the Stop button, the supply to the coil is broken, the coil ceases to be a magnet and the 'hold in' contact returns to its original position (position 1). Since the Start button had already returned to its original open position when we released it, when we take our finger off the Stop button everything will be as we first started. Therefore any loss of supply will immediately break the supply to the coil and stop the motor: if a supply failure was restored, the equipment could not restart itself – someone would have to take the conscious decision to do so.

Figure 7.36 Restarting the motor

This system is basically the same as that of a contactor. In fact, many people refer to this item as the contactor starter. Such starters are also available with a 400 V coil, which is therefore connected across two of the phases.

Remote stop/start control

In the DOL starter as described, we have the means of stopping and starting the motor from the buttons provided on the starter enclosure. However, there are situations where the control of the motor needs to take place from some remote location. This could be, for example, in a college workshop where, in the case of an emergency, emergency stop buttons located throughout the workshop can be activated to break the supply to a motor. Equally, because of the immediate environment around the motor, it may be necessary to operate it from a different location.

Commonly known as a remote stop/start station, the enclosure usually houses a start and a stop button connected in series. However, depending on the circumstances it is also possible to have an additional button included to give 'inch' control of a motor.

If we use the example of our DOL starter as described in the previous diagrams, but now include the remote stop/start station, the circuit would look as in Figure 7.37 below, where for ease the additional circuitry has been shown in red.

Figure 7.37   Remote stop/start control

As can be seen, the remote start button is effectively in parallel with the start button on the main enclosure with the supply to both of these buttons being routed via the stop button of the remote station.

If the intention is to provide only emergency stops, omit the remote station shown so that these are all connected in series with the stop button on the main enclosure.

Hand-operated star-delta starter

This is a two-position method of starting a three-phase squirrel-cage motor, in which the windings are connected firstly in star for acceleration of the rotor from standstill, and then secondly in delta for normal running.

The connections of the motor must be arranged for star-delta, starting with both ends of each phase winding – six in all – brought out to the motor terminal block. The starter itself, in its simplest form, is in effect a changeover switch. Figure 7.38 gives the elementary connections for both star and delta.

Remember

Although hand-operated star-delta starters are now rare, you may still come across them in your career as an electrician

Figure 7.38 Hand-operated star-delta connections

When the motor windings are connected in star, the voltage applied to each phase winding is reduced to 58 per cent of the line voltage, thus the current in the winding is correspondingly reduced to 58 per cent of the normal starting value.

Applying these reduced values to the typical three-phase squirrel-cage induction motor, we would have: initial starting current from two to three-and-a-half times full-load current and initial starting torque of about 50 per cent of the full-load value.

The changeover from star to delta should be made when the motor has reached a steady speed on star connection, at which point the windings will now receive full line voltage and draw full-rated current.

If the operator does not move the handle quickly from the start to run position, the motor may be disconnected from the supply long enough for the motor speed to fall considerably. When the handle is eventually put into the run position, the motor will therefore take a large current before accelerating up to speed again. This surge current could be large enough to cause a noticeable voltage dip. To prevent this, a mechanical interlock is fitted to the operating handle. However, in reality the handle must be moved quickly from start to run position, otherwise the interlock jams the handle in the start position.

The advantage of this type of starter is that it is relatively cheap. It is best suited for motors against no load or light loads, and it also incorporates no-volts protection and overload protection.

Star-delta starter

Automatic star-delta starter

Bearing in mind the user actions required of the previous hand-operated starter, the fully automatic star-delta contactor starter (as shown in Figure 7.39) is the most satisfactory method of starting a three-phase cage-induction motor. The starter consists of three triple-pole contactors, one employing thermal overload protection, the second having a built-in time-delay device, and the third providing the star points.

The changeover from star to delta is carried out automatically by the timing device, which can be adjusted to achieve the best results for a particular situation.

Figure 7.39   Automatic star-delta contactor starter

Soft starters

A soft starter is a type of reduced-voltage starter that reduces the starting torque for a.c. induction motors. The soft starter is in series with the supply to the motor, and uses solid-state devices to control the current flow and therefore the voltage applied to the motor. In theory, soft starters can be connected in series with the line voltage applied to the motor, or can be connected inside the delta loop of a delta-connected motor, controlling the voltage applied to each winding. Soft starters can also have a soft-stop function, which is the exact opposite to soft start, and sees the voltage gradually reduced and thus a reduction in the torque capacity of the motor.

The auto-transformer starter

This method of starting is used when star-delta starting is unsuitable, either because the starting torque would not be sufficiently high using that type of starter, or because only three terminals have been provided at the motor terminal box – a practice commonly found within the UK water industry.

This again is a two-stage method of starting three-phase squirrel-cage induction motors, in which a reduced voltage is applied to the stator windings to give a reduced current at start. The reduced voltage is obtained by means of a three-phase auto transformer, the tapped windings of which are designed to give 40 per cent, 60 per cent and 75 per cent of the line voltage respectively. Although there are a number of tappings, only one tapping is used for the initial starting, as the reduced voltage will also result in reduced torque. Once this has been selected for the particular situation in which the motor is operating, it is left at that position and the motor is started in stages – much like the star-delta starter in that once the motor has reached sufficient speed the changeover switch moves onto the run connections, thus connecting the motor directly to the three-phase supply. Figure 7.40 illustrates the connections for an auto-transformer starter.

Figure 7.40 Connections for an auto-transformer starter

The rotor-resistance starter

This type of starter is used with a slip-ring wound-rotor motor. These motors and starters are primarily used where the motor will start against full load, as an external resistance is connected to the rotor windings through slip rings and brushes, which serves to increase the starting torque.

When the motor is first switched on, the external rotor resistance is at a maximum. As the motor speed increases, the resistance is reduced until at full speed, when the external resistance is completely eliminated and the machine runs as a squirrel-cage induction motor.

The starter is provided with no-volts and overload protection and an interlock to prevent the machine being switched on with no rotor resistance connected. (For clarity these are not shown in Figure 7.41, since the purpose of the diagram is to show the principle of operation.)

Figure 7.41 Rotor-resistance starter

Motor speed control

Speed control of d.c. machines

We said at the beginning of this chapter that there are three types of d.c. motor: series, shunt and compound, and that one of the advantages of the d.c. machine is the ease with which the speed may be controlled.

Some of the more common methods used to achieve speed control on a d.c. machine are described below.

Armature resistance control

With this system of control we are effectively reducing the voltage across the armature terminals by inserting a variable resistor into the armature circuit of the motor. In effect we are creating the illusion of applying a lower-than-rated voltage across the armature terminals.

The disadvantages of this method of control are that we see much of the input energy dissipated in the variable resistor, a loss of efficiency in the motor and poor speed regulation in the shunt and compound motors.

Although not discussed in this book, the principle of applying a lower-than-rated voltage across the armature terminals forms the basis of the Ward-Leonard system of speed control, which in essence provides a variable voltage to the armature terminals by controlling the field winding of a separate generator. Although expensive, this method gives excellent speed control and therefore finds use in situations such as passenger lifts.

Field control

This method works on the principle of controlling the magnetic flux in the field winding. This can be controlled by the field current and as a result controls the motor speed. As the field current is small, the power dissipated by the variable resistor is reasonably small. We can control the field current in the various types of motor as follows:

- **Series motor** – Place a variable resistor in parallel with the series field winding.
- **Shunt motor** – Place a variable resistor in series with the field shunt winding.

This method of speed control is not felt to be suitable for compound machines, as any reduction in the flux of the shunt is offset by an increase in flux from the series field because of an increase in armature current.

Pulse width modulation (PWM)

We know all about the problem with the variable resistor: although it works well, it generates heat and hence wastes power. PWM d.c. motor control uses electronics to eliminate this problem. It controls the motor speed by driving the motor with short pulses. These vary in duration to change the speed of the motor. The longer the pulses, the faster the motor turns, and vice versa. The main disadvantages of PWM circuits are their added complexity and the possibility of generating radio frequency interference (RFI), although this can be minimised by the use of short leads or additional filtering on the power supply leads.

Speed control of a.c. induction motors

We have already established that synchronous speed is directly proportional to the frequency of the supply and inversely proportional to the number of pole pairs. Therefore the speed of an induction motor can be changed by varying the frequency and/or the number of poles. We can also control the speed by changing the applied voltage and the armature resistance.

For adjustable speed applications, variable speed drives (VSD) use these principles by controlling the voltage and frequency delivered to the motor. This gives control over motor torque and reduces the current level during starting. Such drives can control the speed of the motor at any time during operation.

The phrase 'a.c. drive' has different meanings to different people. To some it means a collection of mechanical and electro-mechanical components (the variable frequency inverter and motor combination), which, when connected together, will move a load. More commonly – and also for our purposes – an a.c. drive should be considered as being a variable-frequency inverter unit (the drive) with the motor as a separate component. Manufacturers of variable-frequency inverters will normally refer to these units as being a variable-frequency drive (VFD).

A variable-frequency drive is a piece of equipment that has an input section – the converter – which contains diodes arranged in a bridge configuration. This converts the a.c. supply to d.c. The next section of the VFD, known as the constant-voltage d.c. bus, takes the d.c. voltage and filters and smoothes out the wave form. The smoother the d.c. wave form is, the cleaner the output wave form from the drive.

The d.c. bus then feeds the final section of the drive, the inverter, and as we know from reading earlier chapters of this book, this section of the drive will invert the d.c. back to a.c. using Insulated Gate Bipolar Transistors (IGBT), which create a variable a.c. voltage and frequency output.

On the job: Motor noise

You are asked to investigate a problem in the plant room of a hospital, as the estates maintenance manager has reported a noisy motor on the air-conditioning system. When you look at the motor, you find that as well as it being noisy there is a vibration.

1. Do you think this is a real problem? If so, what actions would you recommend?

(A suggested answer can be found on page 316)

Knowledge check

1. What will be the percentage slip of an induction motor having a synchronous speed of 1,500 rpm and a rotor speed of 1,425 rpm?

2. What is the type of motor most likely to be used in a domestic vacuum cleaner?

3. Describe the principle of operation of a series-wound (universal) motor.

4. How would you change the direction of rotation of a three-phase induction motor?

5. What are the component parts of a squirrel-cage rotor?

6. Describe how to reverse the direction of a single-phase split-phase motor.

7. What is the purpose of the centrifugal switch in a capacitor start motor?

8. What type of motor uses the rotor-resistance starter?

9. In Fleming's left-hand (motor) rule, the second finger is used to represent which quantity?

Chapter 8

Electronics

OVERVIEW

In today's modern technical world electricians can no longer say they do not need to learn about electronics because the use of devices such as security alarms, telephones, dimmers, boiler controls and speed controllers have now brought electronics into general electrical installation work. This chapter will cover:

- **Resistors**
- **Capacitors**
- **Semiconductor devices**
- **Rectification**
- **Transistors**
- **Integrated circuits (ICs)**
- **Thyristors**
- **Field effect transistors (FETs)**

Resistors

There are two basic types of resistor: fixed and variable. The resistance value of a fixed resistor cannot be changed by mechanical means (though its normal value can be affected by temperature or other effects). Variable resistors have some means of adjustment (usually a spindle or slider). The method of construction, specifications and features of both fixed and variable resistor types vary, depending on what they are to be used for.

Fixed resistors

Making a resistor simply consists of taking some material of a known resistivity and making the dimensions (csa and length) of a piece of that material such that the resistance between the two points at which leads are attached (for connecting into a circuit) is the value required.

Most of the very earliest resistors were made by taking a length of resistance wire (wire made from a metal with a relatively high resistivity, such as brass) and winding this on to a support rod of insulating material. The resistance value of the resulting resistor depended on the length of the wire used and its cross sectional area.

This method is still used today, though it has been somewhat refined. For example, the resistance wire is usually covered with some form of enamel glazing or ceramic material to protect it from the atmosphere and mechanical damage. The external and internal view of a typical wire wound resistor is shown in Figure 8.1.

Wire wound resistor

End cap   Vitreous enamel coating

Resistance winding on ceramic former

Figure 8.1 Typical wire wound resistor

Most wire wound resistors can operate at fairly high temperatures without suffering damage, so they are useful in applications where some power may be dissipated. They are, however, relatively difficult to mass-produce, which makes them expensive. Techniques for making resistors from materials other than wire have now been developed for low power applications.

Metal oxide and carbon-composition light-dependent resistors

Resistor manufacture advanced considerably when techniques were developed for coating an insulating rod (usually ceramic or glass) with a thin film of resistive material (see Figure 8.2). The resistive materials in common use today are carbon and metal oxides. Metal end caps fitted with leads are pushed over the ends of the coated rod and the whole assembly is coated with several layers of very tough varnish or similar material to protect the film from the atmosphere and from knocks during handling. These resistors can be mass-produced with great precision at very low cost.

Figure 8.2   Resistor construction

Variable resistors

The development of the techniques for manufacturing variable resistors followed fairly closely that of fixed resistors, though they required some sort of sliding contact together with a fixed resistor element.

Wire wound variable resistors are often made by winding resistance wire onto a flat strip of insulating material, which is then wrapped into a nearly complete circle. A sliding contact arm is made to run in contact with the turns of wire as they wrap over the edge of the wire strip as in Figure 8.3 below. Straight versions are also possible. A straight former is used and the wiper travels in a straight line along it as shown in Figure 8.4.

While wire wound resistors are ideal for certain applications, there are many others where their size, cost and other disadvantages make them unattractive, and as a consequence alternative types have been developed.

The early alternative to the wire wound construction was to make the resistive element (on which the wiper rubs) out of a carbon composition, deposited or moulded as a track and shaped as a nearly completed circle on an insulating support plate. Alternative materials for the track are carbon films, or metal alloys of a metal oxide and a ceramic (cermet) and again straight versions are possible.

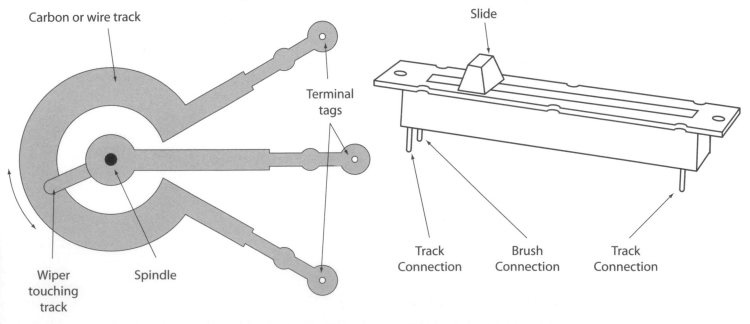

Figure 8.3   Layout of internal track of rotary variable resistor

Figure 8.4   Linear variable resistor

Preferred values

In theory, there's no reason why you couldn't have resistors in every imaginable resistance value; from zero to, say, tens or hundreds of megohms. In reality, however, such an enormous range would be totally impractical to manufacture and store. And from the point of view of the circuit designer it's not usually necessary, either.

So, rather than an overwhelming number of individual resistance values, what manufacturers do is make a limited range of preferred resistance values. In electronics, we use the preferred value closest to the actual value we need.

A resistor with a preferred value of 1000 Ω and a 10 per cent tolerance can have any value between 900 Ω and 1100 Ω. The next largest preferred value, which would give the maximum possible range of resistance values without too much overlap, is 1200 Ω. This can have a value between 1080 Ω and 1320 Ω.

Together, these two preferred value resistors cover all possible resistance values between 900 Ω and 1320 Ω. The next preferred values would be 1460 Ω, 1785 Ω etc.

There is a series of preferred values for each tolerance level as shown in Table 8.1, so that every possible numerical value is covered.

E6 series 20% Tol	E12 series 10% Tol	E24 series 5% Tol	E6 series 20% Tol	E12 series 10% Tol	E24 series 5% Tol
10	10	10	33	33	33
		11			36
	12	12		39	39
		13			43
15	15	15	47	47	47
		16			51
	18	18		56	56
		20			62
22	22	22	68	68	68
		24			75
	27	27		82	82
		30			91

Table 8.1   Table of preferred values

Resistance markings

There is obviously the need for the resistor manufacturer to provide some sort of markings on each resistor so that it can be identified.

The user should be able to tell, by looking at the resistor, what its nominal resistance value is and its tolerance. Various methods of marking this information onto each resistor have been used and sometimes a resistor code will use numbers and letters rather than colours.

Where physical size permits, putting the actual value on the resistor in figures and letters has an obvious advantage in terms of easy interpretation. However, again because of size restrictions, we don't use the actual words and instead use a code system. This code is necessary because when using small text on a small object, certain symbols and the decimal point become very hard to see. This code system is also commonly used to represent resistance values on circuit diagrams for the same reason.

In reality resistance values are generally given in either Ω, kΩ or MΩ using numbers from 1–999 as a prefix (e.g. 10 Ω, 567 kΩ etc.). In the code system we replace Ω, kΩ and MΩ and represent them instead by using the following letters:

- Ω = R
- kΩ = K
- MΩ = M.

These letters are now inserted wherever the decimal point would have been in the value. So for example a resistor of value 10 Ω resistor would now be shown as 10R, and a resistor of value 567 kΩ resistor would become 567 K.

Table 8.2 gives some more examples of this code system. Table 8.3 shows the letters that are then commonly used to represent the tolerance values. These letters are added at the end of the resistor marking so that, for example, a resistor of value 2.7 MΩ with a tolerance of ±10% would be shown as 2M7K.

> **Remember**
>
> Whole numbers could have a decimal point at the end (e.g. 10.0 or 567.0), but we normally miss them out when we write the numbers down (e.g. 10 or 567)

0.1 Ω	is coded	R10
0.22 Ω	is coded	R22
1.0 Ω	is coded	1R0
3.3 Ω	is coded	3R3
15 Ω	is coded	15R
390 Ω	is coded	390R
1.8 Ω	is coded	1R8
47 Ω	is coded	47R
820 kΩ	is coded	820K
2.7 MΩ	is coded	2M7

Table 8.2 Examples of resistance value codes

F	=	± 1%
G	=	± 2%
J	=	± 5%
K	=	± 10%
M	=	± 20%
N	=	± 30%

Table 8.3 Codes for common tolerance values

Remember

Before you read a resistor, turn it so that the end with bands is on the left hand side. Now you read the bands from left to right (as shown in Figures 8.5 to 8.8)

Resistor coding

Standard colour code

Many resistors are so small that it is impractical to print their value on them. Instead, they are marked with a code that uses bands of colour. Located at one end of the component, it is these bands that identify the resistor's value and tolerance.

Most general resistors have four bands of colour, but high precision resistors are often marked with a five-colour band system. No matter which system is being used, the value of the colours is the same.

Resistor colour code

Band colour	Value
Black	0
Brown	1
Red	2
Orange	3
Yellow	4
Green	5
Blue	6
Violet	7
Grey	8
White	9
Gold	0.1
Silver	0.01

Figure 8.5   Resistor colour code

Tolerance colour code

Band colour	±%
Brown	1
Red	2
Gold	5
Silver	10
None	20

Figure 8.6   Tolerance colour code

What this means

Band 1 First figure of value
Band 2 Second figure of value
Band 3 Number of zeros/multiplier
Band 4 Tolerance (±%) See below

Note that the bands are closer to one end than the other

Figure 8.7   What this means

Brown	Green	Orange	Gold
1	5	000	5%

Resistor is 15000Ω or 15K ± 5%

Yellow	Violet	Silver	Gold
4	7	×0.01	5%

Resistor is 47 × 0.01Ω or 0.47K ± 2%

Red	Red	Green	
2	2	00000	20%

Resistor is 2200000Ω or 2.2M ± 20%

Brown	Green	Red	Gold
1	5	00	5%

Resistor is 1500Ω or 1.5K ± 5%

Figure 8.8   Examples of colour coding

Example 1

A resistor is colour coded red, yellow, orange, gold. Determine the value of the resistor.

- first band red (First digit) 2
- second band yellow (Second digit) 4
- third band orange (No of zeros) 3
- fourth band gold (Tolerance) 5%.

The value is 24,000 Ω ±5%.

Example 2

A resistor is colour coded yellow, yellow, blue, silver. Determine the value of the resistor.

- first band yellow (First digit) 4
- second band yellow (Second digit) 4
- third band blue (No of zeros) 6
- fourth band silver (Tolerance) 10%.

The value is 44,000,000 Ω ±10%.

Example 3

A resistor is colour coded violet, orange, brown, gold. Determine the value of the resistor.

- first band violet (First digit) 7
- second band orange (Second digit) 3
- third band brown (No of zeros) 1
- fourth band gold (Tolerance) 5%.

The value is 730 Ω ± 5%.

Example 4

A resistor is colour coded green, red, yellow, silver. Determine the value of the resistor.

- first band green (First digit) 5
- second band red (Second digit) 2
- third band yellow (No of zeros) 4
- fourth band silver (Tolerance) 10%.

The value is 520,000 Ω ±10%.

Remember

To help you remember the resistor colour, learn this rhyme:

Barbara	Black	0
Brown	Brown	1
Runs	Red	2
Over	Orange	3
Your	Yellow	4
Garden	Green	5
But	Blue	6
Violet	Violet	7
Grey	Grey	8
Won't	White	9

Testing resistors

Resistors must be removed from a circuit before testing, otherwise readings will be false. To measure the resistance, the leads of a suitable ohmmeter should be connected to each resistor connection lead and a reading obtained which should be close to the preferred value and within the tolerance stated.

Resistors as current limiters

A resistor is often provided in a circuit to limit, restrict or reduce the current flowing in the circuit to some level that better suits the ratings of some other component in the circuit. For example, consider the problem of operating a solenoid valve from a 36 V d.c. supply, given the information that the energising current of the coil fitted to the valve is 100 mA and its resistance is 240 Ω.

Note that the coil, being a wound component, is actually an inductor. However, we are concerned here with the steady d.c. current through the coil and not the variation in coil current at the instant the supply is connected, so we can ignore the effects of its inductance and consider only the effects of its resistance.

If the solenoid valve were connected directly across the 36 V supply, as shown in Figure 8.9, then from Ohm's law the steady current through its coil would be:

$$I = \frac{V}{R}$$

$$= \frac{36}{240}$$

$$= 0.15 \text{ A or } 150 \text{ mA}$$

Figure 8.9 Solenoid valve connected across 36 V supply

As the coil was designed to produce an adequate magnetic 'pull' when energised at 100 mA, any increase in the energising current is unnecessary and may in fact be highly undesirable due to the resulting increase in the power, which would be dissipated (as heat) within the coils.

Note: The power dissipated in the coil of the solenoid valve, when energised at the recommended current of 100 mA is:

$$P = I^2 \times R$$

$$= 0.1^2 \text{A} \times 240 \,\Omega = 2.4 \text{ watts}$$

Now if the current were to be 0.15 A on the 36 V supply it would be:

$$P = V \times I$$

$$= 36 \times 0.15$$

$$= 5.4 \text{ watts}$$

Thus connecting the coil directly across a 36 volt supply would result in the power dissipation in it being more than doubled. If the valve is required to be energised for more than very brief periods of time, the coil could be damaged by overheating.

Some extra resistance must therefore be introduced into the circuit so that the current through the coil is limited to 100 mA even though the supply is 36 V.

For a current of 100 mA to flow from a 36 V supply, the total resistance R_t connected across the 36 V must, from Ohm's law, be:

$$R_t = \frac{36}{0.1}$$
$$= 360\,\Omega$$

of which there is already 240 Ω in the coil.

A resistor of value 120 Ω must therefore be fitted in series with the coil to bring the value of R_t to 360 Ω. This limits the current through the coil to 100 mA when the series combination of coil and resistor is connected across the 36 V supply as shown in Figure 8.10.

Figure 8.10   Series combination of coil and resistor connected across 36 V supply

Resistors for voltage control

Within a circuit it is often necessary to have different voltages at different stages and we can achieve this by using resistors.

For example, if we physically opened up a resistor and connected its ends across a supply, we would find that, if we then measured the voltage at different points along the resistor, the values would vary along its length. In doing so, we are effectively imitating the 'tapping' technique that we used in the transformers section of this book.

However, reality will stop us from doing this as resistors are sealed components. But we can create the same tapping effect by combining two resistors in series as shown in Figure 8.11, and then our tapping becomes a connection point made between the two resistors.

If we look at Figure 8.11, we can see that the series combination of resistors R_1 and R_2 is connected across a supply that is provided

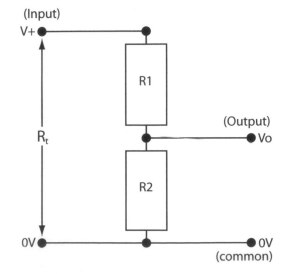

Figure 8.11   Series circuit for voltage control

by two rails. One is shown as V+ (the positive supply rail or in other words our input) and the other as 0V (or common rail of the circuit).

The total resistance of our network (R_t) will be:

$$R_t = R_1 + R_2$$

We know, using Ohm's law, that $V = I \times R$ and the same current flows through both resistors. Therefore for this network, we can see that $V+ = I \times R_t$, and the voltage dropped across resistor R_2, $V_o = I \times R_2$.

We now have two expressions, one for V+ and one for Vo. We can find out what fraction Vo is of V+ by putting Vo over V+ on the left hand side of an equation and then putting what we said each one is equal to in the corresponding positions on the right hand side. This gives us the following formula:

$$\frac{Vo}{V+} = \frac{I \times R_2}{I \times R_t}$$

As current is common on the right hand side of our formula, they cancel each other out. This leaves us with:

$$\frac{Vo}{V+} = \frac{R_2}{R_t}$$

To establish what Vo (our output voltage) actually is, we can transpose again, which would give us:

$$Vo = \frac{R_2 \times V+}{R_t}$$

Finally, we can replace R_t by what it is actually equal to, and this will give us the means of establishing the value of the individual resistors needed to give a desired output voltage (Vo). By transposition, our final formula now becomes:

$$Vo = \frac{R_2}{R_1 + R_2} \times V+$$

This equation is normally referred to as the potential divider rule.

In reality R_1 and R_2 could each be a combination (series or parallel) of many resistors.

However, as long as each combination is replaced by its equivalent resistance so that the simplified circuit looks like Figure 8.11, then the potential divider rule can be applied.

The potential divider circuit is very useful where the full voltage available is not required at some point in a circuit and, as we have seen, by a suitable choice of resistors in the potential divider, the desired fraction of the input voltage can be produced.

In applications where the fraction produced needs to be varied from time to time, the two resistors are replaced by a variable resistor (also known as a potentiometer, which is often abbreviated to the word pot), which would be connected as shown in Figure 8.12.

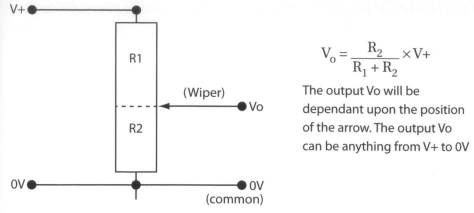

$$V_o = \frac{R_2}{R_1 + R_2} \times V+$$

The output Vo will be dependant upon the position of the arrow. The output Vo can be anything from V+ to 0V

Figure 8.12   Circuit diagram for voltage applied across potentiometer

The potentiometer has a resistor manufactured in the form of a track, the ends of which effectively form our V+ and 0V connections. Our output voltage (Vo) is achieved by means of a movable contact that can touch the track anywhere on its length and this is called the wiper. We have therefore effectively created a variable tapping point.

To compare this with our potential divider, we can say that the part of the track above the wiper can be regarded as R_1 and that part below the wiper as R_2. The fraction of the input voltage appearing at the output can therefore be calculated for any setting of the wiper position by using our potential divider equation:

$$V_o = \frac{R_2}{R_1 + R_2} \times V+$$

Obviously, when the wiper is at the top of the track, R_1 becomes zero and the equation would give the result that Vo is equal to V+. Equally, with the wiper right at the bottom of the track, R_2 now becomes zero and therefore Vo also becomes zero, which is not too surprising as the wiper is now more or less directly connected to the 0V rail.

This sort of circuit finds practical application in a wide variety of control functions such as volume or tone controls on audio equipment, brightness and contrast controls on televisions and shift controls on oscilloscopes.

Power ratings

Resistors often have to carry comparatively large values of current so they must be capable of doing this without overheating and causing damage. As the current has to be related to the voltage, it is the power rating of the resistor that needs to be identified.

The power rating of a resistor is thus really a convenient way of stating the maximum temperature at which the resistor is designed to operate without damage to itself. In general the more power a resistor is designed to be capable of dissipating, the larger physically the resistor is; the resulting larger surface area aids heat dissipation.

Resistors with high power ratings may even be jacketed in a metal casing provided with cooling ribs and designed to be bolted flat to a metal surface – all to improve the radiation and conduction of heat away from the resistance element.

Power is calculated by:

$$P = V \times I$$

Instead of V we can substitute $I \times R$ for V and $\dfrac{V}{R}$ for I. We can then use the following equations to calculate power:

$$P = I^2 \times R$$

Or:

$$P = \dfrac{V^2}{R}$$

What would the power rating of the 50 Ω resistor in Figure 8.13 be?

$P = V \times I$	$= 4 \times 0.08$	$= 0.32$ watts
$P = I^2 \times R$	$= 0.08^2 \times 50$	$= 0.32$ watts
$P = \dfrac{V^2}{R}$	$= \dfrac{4 \times 4}{50}$	$= 0.32$ watts

Figure 8.13 Typical power ratings for resistors

Normally only one calculation is required. Typical power ratings for resistors are shown in Table 8.4.

Carbon resistors	0 to 0.5 watts
Ceramic resistors	0 to 6 watts
Wire wound resistors	0 to 25 watts

Table 8.4 Typical power ratings for resistors

Manufacturers also always quote a maximum voltage rating for their resistors on their data sheets. The maximum voltage rating is basically a statement about the electrical insulation properties of those parts of the resistor, which are supposed to be insulators (e.g. the ceramic or glass rod which supports the resistance element or the surface coating over the resistance element).

If the maximum voltage rating is exceeded there is a danger that a flashover may occur from one end of the resistor to the other. This flashover usually has disastrous results. If it occurs down the outside of the resistor it can destroy not only the protective coating but, on film resistors, the resistor film as well.

If it occurs down the inside of the resistor the ceramic or glass rod is frequently cracked (if not shattered) and, of course, this mechanical damage to the support for the resistance element results in the element itself being damaged as well.

Thermistors

A thermistor is a resistor which is temperature sensitive. The general appearance is shown in the photograph. They can be supplied in various shapes and are used for the measurement and control of temperature up to their maximum useful temperature limit of about 300°C. They are very sensitive and because of their small construction they are useful for measuring temperatures in inaccessible places.

Thermistors are used for measuring the temperature of motor windings and sensing overloads. The thermistor can be wired into the control circuit so that it automatically cuts the supply to the motor when the motor windings overheat, thus preventing damage to the windings.

Thermistors can have a temperature coefficient that may be positive (PTC) or negative (NTC). With a PTC thermistor the resistance of the thermistor increases as the surrounding temperature increases. With the NTC thermistor the resistance decreases as the temperature increases.

The rated resistance of a thermistor may be identified by a standard colour code or by a single body colour used only for thermistors. Typical values are shown in Table 8.5.

Did you know?

Thermistors are used for monitoring the temperature of the water in a motor car

Colour	Resistance
Red	3,000 Ω
Orange	5,000 Ω
Yellow	10,000 Ω
Green	30,000 Ω
Violet	100,000 Ω

Table 8.5   Colour coding for rated resistance of thermistor

Thermistors

Light-dependent resistors

These resistors are sensitive to light. They consist of a clear window with a cadmium sulphide film under it. When light shines onto the film its resistance varies, with the resistance reducing as the light increases.

These resistors are commonly found in street lighting. You may sometimes observe street lights switching on during a thunderstorm in the daytime. This is because the sunlight is obscured by the dark thunderclouds, thus increasing the resistance, which in turn controls the light 'on' circuit.

Light-dependent resistor

Capacitors

Just as resistors enable us to introduce known amounts of resistance into a circuit to serve our purposes, so we can use components known as capacitors to introduce capacitance into a circuit. Like resistance, capacitance always exists in circuits – though, as you'll see when we have discussed the subject in more detail, capacitance exists between conductors whereas resistance exists in conductors.

Basic principles

A capacitor is basically two metallic surfaces usually referred to as plates, separated by an insulator commonly known as the dielectric. The plates are usually, though not necessarily, metal and the dielectric is any insulating material. Air, glass, ceramic, mica, paper, oils and waxes are some of the many materials commonly used. The common symbols used for capacitors are identified in Figure 8.14.

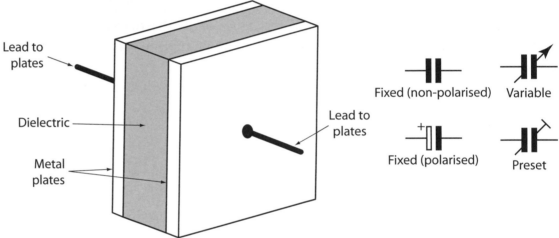

Figure 8.14   Basic construction of a capacitor and circuit symbols

These two plates are not in contact with each other and as such they do not form a circuit in the same way that conductors with resistors do. However, the capacitor stores a small amount of electric charge and as such it can be thought of as a small rechargeable battery, which can be quickly recharged.

The capacitance of any capacitor depends on three factors:

1. It depends on the working area of the plates i.e. the area of the conducting surfaces facing each other. We can think of the degree of crowding of excess electrons near the surface of one plate of a capacitor (and the corresponding sparseness of electrons near the surface of the other) as being directly related to the potential difference (p.d.) applied across the capacitor, for example connecting it directly across a battery. If we increase the area of the plates, then more electrons can flow onto one of the plates before the same degree of crowding is reached. The battery voltage determines this level of crowdedness. There is of course a similar increased loss of electrons from the other plate. The working area of the plates is directly proportional to the capacitance. If we double the area of the plates we double the capacitance of the capacitor.

Remember

In any application the capacitor to be used should meet or preferably exceed the capacitor voltage rating. The voltage rating is often called 'the working voltage' and refers to d.c. voltage values. When applied to a.c. circuits the peak voltage value must be used as a comparison to the d.c. working voltage of a capacitor

2. Capacitance depends on the thickness of the **dielectric** between the plates. As was mentioned earlier, the capacitance effect depends on the forces of repulsion or attraction caused by an electron surplus or shortage on the plates on either side of the dielectric. The further apart the plates are, the weaker these factors become. As a result, the degree of crowding of electrons on one plate (and the shortage of electrons on the other) produced by a given p.d across the capacitor decreases.

3. The capacitance depends on the nature of the dielectric or spacing material used. This fundamental principle of capacitors and the time constant of capacitor resistor circuits will be looked at later under the heading of electrostatics.

Capacitor types

There are two major types of capacitor, fixed and variable, both of which are used in a wide range of electronic devices. Fixed capacitors can be further subdivided into electrolytic and non-electrolytic types and together they represent the majority of the market.

All capacitors possess some resistance and inductance because of the nature of their construction. These undesirable properties result in limitations, which often determine their applications.

Fixed capacitors

Electrolytic capacitors

These capacitors have a much higher capacitance, volume for volume, than any other type. This is achieved by making the plate separation extremely small by using a very thin dielectric (insulator). The dielectric is often mica or paper.

They are constructed on the Swiss roll principle as are the paper dielectric capacitors used for Power Factor correction in electrical installation circuits, for example fluorescent lighting circuits.

The main disadvantage of an electrolytic capacitor is that it is polarised and must be connected to the correct polarity in a circuit, otherwise a short circuit and destruction of the capacitor will result.

Figure 8.15 illustrates a newer type of electrolytic capacitor using tantalum and tantalum oxide to give a further capacitance/size advantage. It looks like a raindrop with two leads protruding from the bottom. The polarity and values may be marked on the capacitor (see photograph in Polarity section on page 280) or the colour code (see Figure 8.22 on page 278) can be used.

Figure 8.15 Tantalum capacitor

Non-electrolytic capacitors

There are many different types of non-electrolytic capacitor. However, only mica, ceramic and polyester are of any significance. Older types using glass and vitreous enamel are expected to disappear over the next few years and even mica will be replaced by film types.

Mica

Mica is a naturally occurring dielectric and has a very high resistance; this gives excellent stability and allows the capacitors to be accurate within a value of ±1 per cent of the marked value. Since costs usually increase with increased accuracy, they tend to be more expensive than plastic film capacitors. They are used where high stability is required, for example in tuned circuits and filters required in radio transmission. Figure 8.16 illustrates a typical mica capacitor.

Figure 8.16   Mica capacitor

Ceramic capacitors

These consist of small rectangular pieces of ceramic with metal electrodes on opposite surfaces. Figure 8.17 illustrates a typical ceramic capacitor. These capacitors are mainly used in high-frequency circuits subjected to wide temperature variations. They have high stability and low loss.

Figure 8.17   Ceramic capacitor

Polyester capacitors

These are an example of a plastic film capacitor. Polypropylene, polycarbonate and polystyrene capacitors are other types of plastic film capacitors. They are widely used in the electronics industry due to their good reliability and relative low cost but are not suitable for high-frequency circuits. Figure 8.18 illustrates a typical polyester capacitor; however, they can also be a tubular shape (see Figure 8.19).

Figure 8.18   Polyester capacitor

Figure 8.19   Tubular capacitor

Variable capacitors

Variable capacitors generally have air or a vacuum as the dielectric, although ceramics are sometimes used. The two main sub-groups are tuning and trimmer capacitors.

Tuning capacitors

These are so called because they are used in radio tuning circuits and consist of two sets of parallel metal plates, one isolated from the mounting frame by ceramic supports while the other is fixed to a shaft which allows one set to be rotated into or out of the first set. The rows of plates interlock like fingers, but do not quite touch each other.

Figure 8.20   A typical variable capacitor of the tuning type

Trimmer capacitors

These are constructed of flat metal leaves separated by a plastic film; these can be screwed towards each other. They have a smaller range of variation than tuning capacitors, and so are only used where a slight change in value is needed.

Figure 8.21   A typical capacitor used as a trimmer

Capacitor coding

To identify a capacitor the following details must be known: the capacitance, working voltage, type of construction and polarity (if any). The identification of capacitors is not easy because of the wide variation in shapes and sizes. In the majority of cases the capacitance will be printed on the body of the capacitor, which often gives a positive identification that the component is a capacitor.

The capacitance value is the farad (symbol F); this was named after the English scientist Michael Faraday. However, for practical purposes the farad is much too large and in electrical installation work and electronics we use fractions of a farad as follows:

- 1 microfarad = $1\,\mu$F = 1×10^{-6} F
- 1 nanofarad = $1\,$nF = 1×10^{-9} F
- 1 picofarad = $1\,$pF = 1×10^{-12} F.

The power factor correction capacitor found in fluorescent luminaries would have a value typically of 8 μF at a working voltage of 400 V. One microfarad is one million times greater than one picofarad.

The working voltage of a capacitor is the maximum voltage that can be applied between the plates of the capacitor without breaking down the dielectric insulating material.

Did you know?

An ideal capacitor, which is isolated, will remain charged forever. But in practice no dielectric insulating material is perfect, and therefore the charge will slowly leak between the plates, gradually discharging the capacitor. The loss of charge by leakage should be very small for a practical capacitor

It was quite common for capacitors to be marked with colour codes but today relatively few capacitors are colour coded. At one time nearly all plastic foil type capacitors were colour coded, as in Figure 8.22, but this method of marking is rarely encountered. However, it is a useful skill to know and be able to use the colour-coding method as shown in Table 8.6.

This method is based on the standard four-band resistor colour coding. The first three bands indicate the value in normal resistor fashion, but the value is in picofarads. To convert this into a value in nanofarads it is merely necessary to divide by 1000. Divide the marked value by 1,000,000 if a value in microfarads is required. The fourth band indicates the tolerance, but the colour coding is different to the resistor equivalent. The fifth band shows the maximum working voltage of the component. Details of this colour coding are shown in Figure 8.22 and Table 8.6.

Plastic film series C280 capacitors

Band (a)
Band (b)
Band (c)
Band (d)
Band (e)

Figure 8.22   Capacitor colour bands

Standard capacitor colour coding

Colour	1st Digit	2nd Digit	3rd Digit	Tol. Band	Max. Voltage
Black		0	None	20%	
Brown	1	1	1		100 V
Red	2	2	2		250 V
Orange	3	3	3		
Yellow	4	4	4		400 V
Green	5	5	5	5%	
Blue	6	6	6		630 V
Violet	7	7	7		
Grey	8	8	8		
White	9	9	9	10%	

Table 8.6   Standard capacitor colour coding

Bands are then read from top to bottom. Digit 1 gives the first number of the component value; the second digit gives the second number. The third band gives the number of zeros to be added after the first two numbers and the fourth band indicates the capacitor tolerance, which is normally black 20%, white 10% and green 5%.

Example 1

A plastic film capacitor is colour coded from top to bottom brown, red, yellow, black, red. Determine the value of the capacitor, its tolerance and working voltage.

- band (a) – brown = 1
- band (b) – red = 2
- band (c) – yellow = 4 multiply by 10,000
- band (d) – black = 20% tolerance
- band (e) – red = 250 volts.

The capacitor has a value of 120,000 pF or 0.12 μF with a tolerance of 20% and a maximum working voltage of 250 volts.

Example 2

A plastic film capacitor is colour coded from top to bottom orange, orange, yellow, green, yellow. Determine the value of the capacitor, its tolerance and working voltage.

- band (a) – orange = 3
- band (b) – orange = 3
- band (c) – yellow = 4 multiply by 10,000
- band (d) – green = 5%
- band (e) – yellow = 400 volts.

The capacitor has a value of 330,000 pF or 0.33 μF with a tolerance of 5% and a maximum working voltage of 400 volts.

Example 3

A plastic film capacitor is colour coded from top to bottom violet, blue, orange, black, and brown. Determine the value of the capacitor, its tolerance and working voltage.

- band (a) – violet = 7
- band (b) – blue = 6
- band (c) – orange = 3 multiply by 1,000
- band (d) – black = 20%
- band (e) – brown = 100 volts.

The capacitor has a value of 76,000 pF or 0.076 μF with a tolerance of 20% and a maximum working voltage of 100 volts.

Often the value of a capacitor is simply written on its body, possibly together with the tolerance and/or its maximum operating voltage. The tolerance rating may be omitted, and it is generally higher for capacitors than resistors. Most modern resistors have tolerances of 5% or better, but for capacitors the tolerance rating is generally 10% or 20%. The tolerance figure is more likely to be marked on a close tolerance capacitor than a normal 10% or 20% type.

The most popular form of value marking on modern capacitors is for the value to be written on the components in some slightly cryptic form. Small ceramic capacitors generally have the value marked in much the same way that the value is written on a circuit diagram.

Where the value includes a decimal point, it is standard practice to use the prefix for the multiplication factor in place of the decimal point. This is the same practice as was used for resistors.

The abbreviation μ means microfarad; n means nanofarad; p means picofarad. Therefore:

- a 3.5 pF capacitor would be abbreviated to 3p5
- a 12 pF capacitor would be abbreviated to 12p
- a 300 pF capacitor would be abbreviated to 300p or n30
- a 4,500 pF capacitor would be abbreviated to 4n5
- 1,000 pF = 1 nF = 0.001 μF.

Polarity

Once the size, type and d.c. voltage rating of a capacitor have been determined it now remains to ensure that its polarity is known. Some capacitors are constructed in such a way that if the component is operated with the wrong polarity its properties as a capacitor will be destroyed, especially electrolytic capacitors. Polarity may be indicated by a + or – as appropriate. Electrolytic capacitors that are contained within metal cans will have the can casing as the negative connection. If there are no markings a slight indentation in the case will indicate the positive end.

Capacitor showing polarity

Tantalum capacitors have a spot on one side as shown in Figure 8.23.

When this spot is facing you the right hand lead will indicate the positive connection.

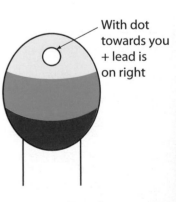

With dot towards you + lead is on right

Figure 8.23 Tantalum capacitor

Electrostatics and calculations with capacitors

The charge stored on a capacitor is dependent on three main factors: the area of the facing plates; the distance between the plates; and the nature of the dielectric. The charge stored by a capacitor is measured in coulombs (Q) and is related to the value of capacitance and the voltage applied to the capacitor:

Charge (coulombs) = Capacitance (farads) × Voltage (volts)

$$Q = C \times V$$

The formula for energy stored in a capacitor can be calculated by using the formula:

$$W = \frac{1}{2}CV^2$$

Capacitors in combination

Capacitors, like resistors, may be joined together in various combinations of series or parallel connections. Figures 8.24 and 8.25 illustrate the equivalent capacitance C_t of a number of capacitors. C_t can be found by applying similar formulae as for resistors. However, these formulae are the opposite way round to series and parallel resistors.

$$C_t = C_1 + C_2 \text{ (as in series resistance)}$$

Figure 8.24 Capacitors connected in parallel

$$\frac{1}{C_t} = \frac{1}{C_1} + \frac{1}{C_2} \text{ (as in parallel resistance)}$$

or

$$C_t = \frac{C_1 \times C_2}{C_1 + C_2} \text{ (when there are two capacitors in series)}$$

Figure 8.25 Capacitors connected in series

Example 1

Capacitors of 10 μF and 40 μF are connected in series and then in parallel. Calculate the effective capacitance for each connection.

Series:

$$\frac{1}{C_t} = \frac{1}{C_1} + \frac{1}{C_2}$$

$$\frac{1}{C_t} = \frac{1}{10\,\mu F} + \frac{1}{40\,\mu F}$$

$$\frac{1}{C_t} = \frac{4\,\mu F + 1\,\mu F}{40\,\mu F}$$

$$\frac{1}{C_t} = \frac{5\,\mu F}{40\,\mu F}$$

Parallel:

$$C_t = C_1 + C_2$$

$$C_t = 10\,\mu F + 40\,\mu F$$

$$C_t = 50\,\mu F$$

Therefore:

$$\frac{C_t}{1\,\mu F} = \frac{40\,\mu F}{5\,\mu F}$$

$$C_t = 8\mu F$$

Example 2

Three capacitors of 30 μF, 20 μF and 15 μF are connected in series across a 400 V d.c. supply. Calculate the total capacitance and the charge on each capacitor.

$$\frac{1}{C_t} = \frac{1}{C_1} + \frac{1}{C_2} + \frac{1}{C_3}$$

$$= \frac{1}{30\ \mu F} + \frac{1}{20\ \mu F} + \frac{1}{15\ \mu F}$$

$$= \frac{9\ \mu F}{60\ \mu F}$$

$$\therefore C_t = 6.66\ \mu F$$

Q, the charge, is common to each capacitor. Therefore:

$$Q = C \times V$$

$$Q = 6.66 \times 10^{-6} \times 400$$

$$Q = 2.664\ m\ coulombs$$

Example 3

Three capacitors of 30 μF, 20 μF and 15 μF are connected in parallel across a 400 V d.c. supply. Calculate the total capacitance, the total charge and the charge on each capacitor.

$$C_t = C_1 + C_2 + C_3$$

$$C_t = 30 + 20 + 15$$

$$C_t = 65\ \mu F$$

Total charge $Q = C \times V$

$$Q_t = C_1 \times V$$

$$Q_t = 65 \times 10^{-6} \times 400$$

$$Q_t = 2.6\ m\ coulombs$$

$$Q_1 = C_1 \times V$$

$$Q_1 = 30 \times 10^{-6} \times 400$$

$$Q_1 = 1.2\ m\ coulombs$$

$$Q_2 = C_2 \times V$$

$$Q_2 = 20 \times 10^{-6} \times 400$$

$$Q_2 = 0.8\ m\ coulombs$$

$$Q_3 = C_3 \times V$$

$$Q_3 = 15 \times 10^{-6} \times 400$$

$$Q_3 = 0.06\ m\ coulombs$$

Charging and discharging capacitors

When a capacitor is connected to a battery, positive and negative charges are deposited on the capacitor plates and the capacitor will charge up. These charges build up fast but not instantaneously; they follow a set pattern of charge. A typical charge and discharge circuit is illustrated in Figure 8.26.

Figure 8.26   A typical charge and discharge circuit

When the switch contact is connected to 1, current flows from the battery to the capacitor plates and the electrolytic capacitor will charge up to the same voltage as the battery.

From the time the current (measured by the microammeter) is switched on a charge curve starts that is related to the charge voltage (measured by the voltmeter) and time. The maximum voltage that the capacitor can take is that of the supply; this then sets the voltage scale from 0 volts to maximum voltage (9 V in this example).

The time scale is divided into sections called time constants. The pattern of charge is always the same. Figure 8.27 illustrates a graph of voltage against time for charging the capacitor.

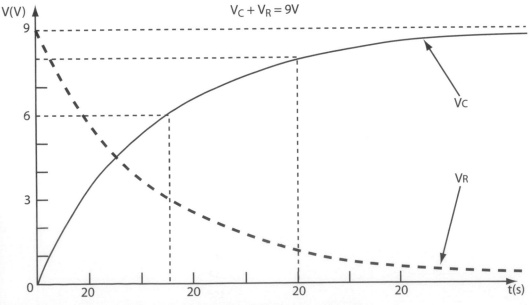

Figure 8.27   Voltage against time graph for charging the capacitor

At 'switch on' the voltage V_C across the capacitor is zero and the voltage V_R across the resistor is at maximum supply voltage.

As the current flow charges up the capacitor plates, V_C increases with the charging of the capacitor plates. V_R reduces over the charge time because the rate of current flow reduces since the supply voltage and the voltage across the capacitor will be at the same potential at completion of charge, therefore no current will be flowing. The current will follow a similar line to V_R on the graph.

The time scale has been subdivided into equal time periods, each time period known as a time constant. In the first time period the voltage reaches approximately two-thirds of the maximum volts (6 V), in the next period the charge goes from the finishing point in the first time constant to finish about two-thirds of what is left (2 V).

This pattern continues until the capacitor is fully charged. The rate of charge will be exponential as shown in Figure 8.28 and it will take five time constants for the capacitor to be fully charged. This also applies if the capacitor is discharging through a series resistor; the voltage (V_C) across the capacitor will fall exponentially and it will again take five time constants for the capacitor to be fully discharged.

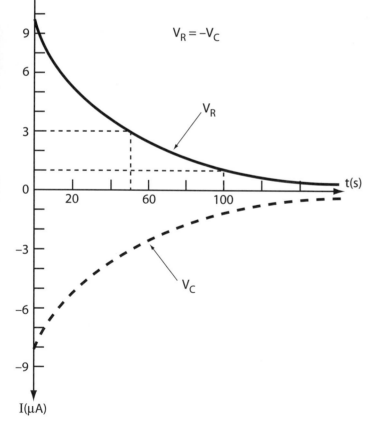

Graph of voltages V_R and V_C

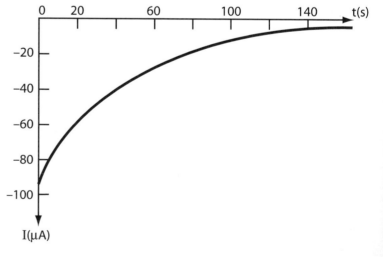

Graph of current I(μA)

Figure 8.28   Discharge curve

Semiconductor devices

The crystal radio that can be constructed from modern electronic game sets depends on the detector action produced by a 'cat's whisker' and a crystal (a cat's whisker was a piece of wire, the point of which was pressed firmly into contact with a suitably mounted piece of natural crystal).

This crystal detector was in fact a diode. It is the use of diodes, semiconductors and semiconductor devices that we will investigate in this section.

Semiconductor basics

Try to think of a semiconductor as being a material that has an electrical quality somewhere between a conductor and an insulator, in that it is neither a good conductor nor a good insulator. Typically, we use semi-conducting materials such as silicon or germanium, where the atoms of these materials are arranged in a 'lattice' structure. The lattice has atoms at regular distances from each other, with each atom 'linked' or 'bonded' to the four atoms surrounding it. Each atom then has four **valence electrons**.

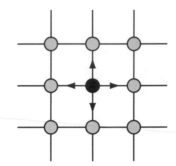

Figure 8.29 Lattice structure of semi-conducting material

However, we have a problem in that, with atoms of pure silicon or germanium no conduction is possible because we have no free electrons. To allow conduction to take place we add an impurity to the material via a process known as **doping**. When we dope the material we can add two types of impurity:

- pentavalent – e.g. arsenic which contains five valence electrons
- trivalent – e.g. aluminium that contains three valence electrons.

As we can see by the number of valence electrons in each, adding a pentavalent (five) material introduces an extra electron to the semiconductor and adding a trivalent (three) material to the semiconductor 'removes' an electron (also known as creating a hole).

When we have an extra electron, we have a surplus of negative charge and call this type of material 'n-type'. When we have 'removed' an electron we have a surplus of positive charge and call this material 'p-type'. It is the use of these two materials that will allow us to introduce the component responsible for rectification, the diode.

The p–n junction

A semiconductor diode is basically created when we bring together an 'n-type' material and a 'p-type' material to form a p–n junction. The two materials form a barrier where they meet which we call the **depletion** layer. In this barrier, the coming together of unlike charges causes a small internal p.d. to exist.

We now need to connect a battery across the ends of the two materials, where we call the end of the p-type material the anode and the end of the n-type material the cathode.

Figure 8.30 p–n junction

If the anode is positive and the battery voltage is big enough, it will overcome the effect of the internal p.d. and push charges (both positive and negative) over the junction. In other words, the junction has a low enough resistance for current to flow. This type of connection is known as being forward biased.

Reverse the battery connections so that the anode is now negative and the junction becomes high resistance and no current can flow. This type of connection is known as being reverse biased.

When the junction is forward biased, it only takes a small voltage (0.7 V for silicon) to overcome the internal barrier p.d.

When reversed biased, it takes a large voltage (1200 V for silicon) to overcome the barrier and thus destroy the diode, effectively allowing current to flow in both directions. As a general summary of its actions, we can therefore say that a diode allows current to flow through it in one direction only.

We normally use the symbol in Figure 8.31 to represent a diode.

In this symbol, the direction of the arrow can be taken to represent the direction of current flow.

Diodes

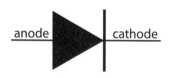

Figure 8.31 Symbol representing a diode

Zener diode

We have just established that a conventional diode will not allow current flow if reverse biased and below its reverse breakdown voltage. We also said that when forward biased (in the direction of the arrow), the diode exhibits a voltage drop of roughly 0.6V for a typical silicon diode. If we were to exceed the breakdown voltage, the internal barrier of the diode would be destroyed, thus allowing current flow in both directions. However, this normally results in the total destruction of the device.

Figure 8.32   Zener diode characteristics

Zener diodes are p–n junction devices that are specifically designed to operate in the reverse breakdown region without completely destroying the device. The breakdown voltage of a zener diode V_z, (known as the zener voltage, named after the American physicist Clarence Zener who first discovered the effect) is set by carefully controlling the doping level during manufacture. The breakdown voltage can be controlled quite accurately in the doping process and tolerances to within 0.05 per cent are available, although the most widely used tolerances are 5 per cent and 10 per cent.

Therefore, a reverse biased zener diode will exhibit a controlled breakdown, allow current to flow and thus keep the voltage across the zener diode at the predetermined zener voltage. Because of this characteristic, the zener diode is commonly used as a form of voltage limiting/regulation when connected in parallel across a load.

When connected so that it is reverse biased in parallel with a variable voltage source, a zener diode acts as a short circuit when the voltage reaches the diode's reverse breakdown voltage and therefore limits the voltage to a known value.

A zener diode used in this way is known as a shunt voltage regulator (shunt meaning connected in parallel and voltage regulator being a class of circuit that produces a fixed voltage).

For a low current power supply, a simple voltage regulator could be made with a resistor (to limit the operating current) and a reverse biased zener diode as shown in Figure 8.33. Here, Vs is the supply voltage, remembering that V_Z is our zener breakdown voltage.

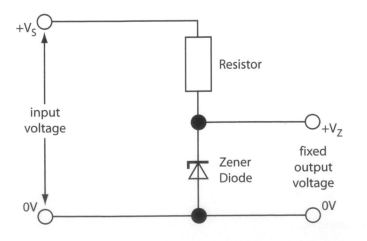

Figure 8.33   A simple voltage regulator made with a resistor and a reverse biased zener diode

As a summary, we can therefore say that a zener's properties are as follows:

- When forward biased (although not normally used for a zener) the behaviour is like an ordinary semiconductor diode.

- When reverse biased, at voltages below V_z the device essentially doesn't conduct, and it behaves just like an ordinary diode.

- When reverse biased, any attempt to apply a voltage greater than V_z causes the device to be prepared to conduct a very large current. This has the effect of limiting the voltage we can apply to around V_z.

As with any characteristic curve, the voltage at any given current, or the current at any given voltage, can be found from the curve of a zener diode.

Did you know?

Zener diodes are readily available with power ratings ranging from a few hundred milliwatts to tens of watts. Low power types are usually encapsulated in glass or plastic packages, and heat transfer away from the junction is mainly by conducting along the wires. High power types like power rectifiers are packaged in metal cases designed to be fitted to heat sinks so that heat can be dissipated by conduction, convection and radiation

Light emitting diodes (LEDs)

The light emitting diode is a p–n junction especially manufactured from a semi-conducting material, which emits light when a current of about 10 mA flows through the junction. No light is emitted if the diode is reverse biased and if the voltage exceeds 5 volts then the diode may be damaged. If the voltage exceeds 2 volts then a series connected resistor may be required.

Figure 8.34 illustrates the general appearance of a LED.

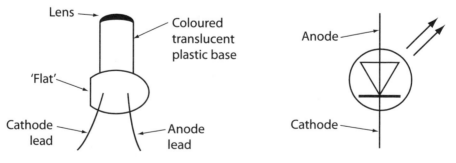

Figure 8.34   Light emitting diode

Photo cell and light-dependent resistor

Figure 8.35   Photo cell and its circuit symbol

The photo cell shown in Figure 8.35 changes light (also infrared and ultraviolet radiation) into electrical signals and is useful in burglar and fire alarms as well as in counting and automatic control systems. Photoconductive cells or light-dependent resistors make use of the semiconductors whose resistance decreases as the intensity of light falling on them increases. The effect is due to the energy of the light setting free electrons from donor atoms in the semiconductor, making it more conductive. The main use of this type of device is for outside lights along the streets, roads and motorways. There are also smaller versions for domestic use within homes and businesses.

Photodiode

A photodiode is a p–n junction designed to be responsive to optical input. As a result, they are provided with either a window or optical fibre connection that allows light to fall on the sensitive part of the device.

Photodiodes can be used in either zero bias or reverse bias. When zero biased, light falling on the diode causes a voltage to develop across the device, leading to a current in the forward bias direction. This is called the photovoltaic effect and is the principle of operation of the solar cell, a solar cell being a large number of photodiodes. The use of the solar cell can then be seen in providing power to equipment such as calculators, solar panels and satellites orbiting the earth

Figure 8.36   Circuit symbol for photodiode

When reverse biased, diodes usually have extremely high resistance. This resistance is reduced when light of an appropriate frequency shines on the junction. When light falls on the junction, the energy from the light breaks down bonds in the 'lattice' structure of the semiconductor material, thus producing electrons and allowing current to flow. Circuits based on this effect are more sensitive to light than ones based on the photovoltaic effect. Consequently, the photodiode is used as a fast counter or in light meters to measure light intensity, as in Figure 8.36.

Opto-coupler

The opto-coupler, also known as an opto-isolator, consists of an LED combined with a photodiode or phototransistor in the same package, as shown in Figure 8.37 and photograph.

Figure 8.37 Opto-coupler circuit

Opto-coupler package

The opto-coupler package allows the transfer of signals, analogue or digital, from one circuit to another in cases where the second circuit cannot be connected electrically to the first, for example, due to different voltages.

Light (or infrared) from the LED falls on the photodiode/transistor which is shielded from outside light. A typical use for one of these is in a VCR to detect the end or start of the tape.

Infrared source and sensor

An infrared beam of light is projected from an LED which is a semiconductor made from gallium arsenide crystal. The light emitted is not visible light, but very close to the white light spectrum. Figure 8.38 shows various housings for both the source output and the sensors within the security alarms industry.

Infrared beams have a receiver, which reacts to the beam in differing ways depending upon its use. Infrared sources/receivers are used for alarm detection and as remote control signals for many applications. The passive infrared (PIR) detector is housed in only one enclosure and uses ceramic infrared detectors. The device does not have a projector but detects the infrared heat radiated from the human body.

Source/Receiver   Source

Receiver   Source/Receiver

Figure 8.38 Housings for infrared source output and sensors for security alarms

Fibre optic link

The simplified block diagram in Figure 8.39 shows a system of communication, which can be several thousand kilometres in length. On the far left we input information such as speech or visual pictures as electrical signals. They are then pulse code modulated in the coder and changed into equivalent digital light signals by the optical transmitter via a miniature laser or LED at the end of the fibre optic cable.

Figure 8.39 Simplified block diagram of fibre optic link

The light is transmitted down the cable to the optical receiver which uses a photodiode or phototransistor which converts the incoming signals back to electrical signals before they are decoded back into legible information.

The advantages of this type of link over a conventional communication system are:

- high information carrying capacity
- free from the noise of electrical interference
- greater distance can be covered, as there is no volt drop
- the cable is lighter, smaller and easier to handle than copper
- crosstalk between adjacent channels is negligible
- it offers greater security to the user.

The fibre optic cable

The fibre optic cable (see Figure 8.40) has a glass core of higher refractive index than the glass cladding around it. This maintains the light beam within the core by total internal reflection at the point where the core and the cladding touch.

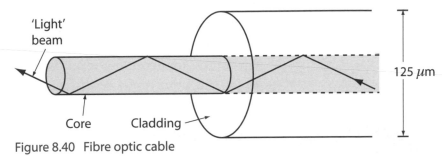

Figure 8.40 Fibre optic cable

This is similar to the insulation of the single core cable preventing the current leaking from the conductor. The beam of light bounces off the outer surface of the core in a zigzag formation along its length.

There are two main types of cable: multimode and singlemode (monomode).

Multimode

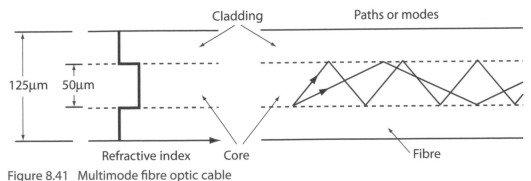

Figure 8.41 Multimode fibre optic cable

The wider core of the multimode fibre (Figure 8.41) allows the infrared to travel by different reflected paths or modes. Paths that cross the core more often are longer and take more time to travel along the fibre. This can sometimes cause errors and loss of information.

Singlemode

The core of the singlemode fibre (Figure 8.42) is about one tenth of the multimode and only straight through transmission is possible.

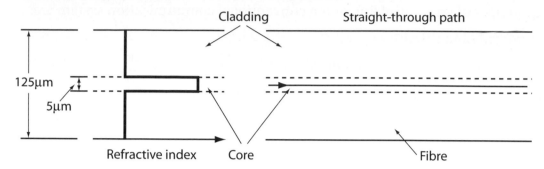

Figure 8.42 Singlemode fibre optic cable

Solid state temperature device (thermistors)

Thermistors monitor the changes in temperature of components, which could become damaged due to excessive heat. As the temperature rises there is a reaction within the semiconductor for its resistance to either rise (positive temp. coefficient) or fall (negative temp. coefficient) depending upon its make up. There are two types of thermistor:

- ptc – made from barium titanate
- ntc – made from oxides of nickel, manganese, copper and cobalt.

We also used **thermocouples**, which are two different metals bonded together and each has a lead. When the bonded metals are heated a voltage appears across the two leads. The hotter the metals become, the larger the voltage (mV). An example of their use is for measuring the temperature of furnaces within the steel industry.

Thermocouples

Diode testing

The p–n junction diode has a low resistance when a voltage is applied in the forward direction and a high resistance when applied in the reverse direction.

Connecting an ohmmeter with the red positive lead to the anode of the junction diode and the black negative lead to the cathode would give a very low reading, reversing the lead connections would give a high resistance reading in a good component.

Figures 8.43 and 8.44 illustrate these connections.

Remember

When using a multi-meter you need to put it into the diode test mode, otherwise when testing on the ohms range you could get a reversed reading

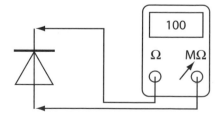

Figure 8.43 High resistance connection

Figure 8.44 Low resistance connection

Rectification

Rectification is the conversion of an a.c. supply into a d.c. supply. Despite the common use of a.c. systems in our day-to-day work as an electrician, there are many applications (e.g. electronic circuits and equipment) that require a d.c. supply. The following section looks at the different forms this can take.

Half-wave rectification

We now know that a diode will only allow current to flow in one direction and it does this when the anode is more positive than the cathode. In the case of an a.c. circuit, this means that only the positive half cycles are allowed 'through' the diode and, as a result, we end up with a signal that resembles a series of 'pulses'. This tends to be unsuitable for most applications, but can be used in situations such as battery charging. A transformer is also commonly used at the supply side to ensure that the output voltage is to the required level. The waveform for this form of rectification would look as in Figure 8.45.

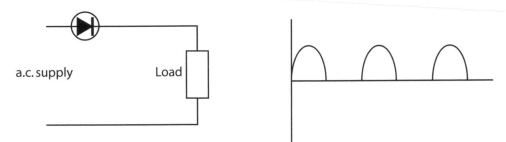

Figure 8.45   Half-wave rectification

Full-wave rectification

We have seen that half-wave rectification occurs when one diode allows the positive half cycles to pass through it. However, we can connect two diodes together to give a more even supply. We call this type of circuit **biphase**. In this method, we connect the anodes of the diodes to the opposite ends of the secondary winding of a centre-tapped transformer. As the anode voltages will be 180° out of phase with each other, one diode will effectively rectify the positive half cycle and one will rectify the negative half cycle. The output current will still appear to be a series of pulses, but they will be much closer together, with the waveform shown in Figure 8.46.

Figure 8.46   Full-wave rectification

The full-wave bridge rectifier

This method of rectification does have to rely on the use of a centre-tapped transformer, but the output waveform will be the same as that of the biphase circuit previously described. In this system, we use four diodes, connected in such a way that at any instant in time two of the four will be conducting. The connections would be as in Figure 8.47, where we have shown two drawings to represent the route through the network for each half cycle.

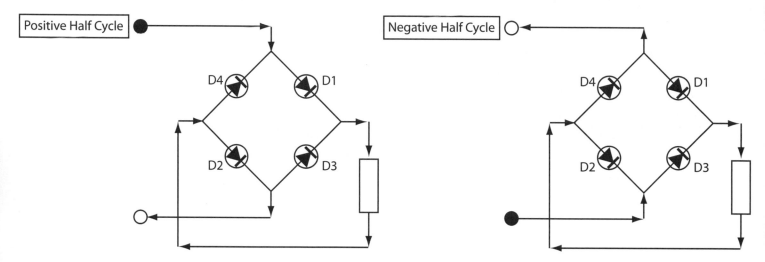

Figure 8.47   Full-wave bridge rectifier

The circuits that we have looked at so far convert a.c. into a supply which, although never going negative, is still not a true d.c. supply. This brings us to the next stage of the story – smoothing.

Smoothing

We have seen that the waveform produced by our circuits so far could best be described as having the appearance of a rough sea. The output current is not at a constant value, but constantly changing. This, as we have said before, is acceptable for battery charging, but useless for electronic circuits where a smooth supply voltage is required.

To make it useful for electronic circuits, we need to smooth out the waveform by creating a situation that is sometimes referred to as **ripple-free** and in essence there are three ways to achieve this, namely capacitor smoothing, choke smoothing and filter circuits.

Capacitor smoothing

If we connect a capacitor in parallel across the load, then the capacitor will charge up when the rectifier allows a flow of current and discharge when the rectifier voltage is less than the capacitor. However, the most effective smoothing comes under no-load conditions. The heavier the load current, the heavier the ripple. This means that the capacitor is only useful as a smoothing device for small output currents.

Choke smoothing

If we connect an inductor in series with our load, then the changing current through the inductor will induce an e.m.f. in opposition to the current that produced it. This means that the e.m.f. will try to maintain a steady current. Unlike the capacitor, this means that the heavier the ripple (rate of change of current) the more that ripple will be smoothed. This effectively means that the choke is more useful in heavy current circuits.

Filter circuits

This is the name given to a circuit that removes the ripple and is basically a combination of the two previous methods. The most effective of these is the capacitor input filter, shown in Figure 8.48.

Figure 8.48   Capacitor input filter

The waveform for the filter circuit that we have just spoken about is shown in Figure 8.49, where the dotted line indicates the waveform before smoothing.

Figure 8.49   Waveform for capacitor input filter

Three-phase rectifier circuits

Whereas Figure 8.49 indicates that a reasonably smooth waveform can be obtained from a single-phase system, we can get a much smoother wave from the three-phase supply mains. To do so we use six diodes connected as a three-phase bridge circuit. These types of rectifier are used to provide high-powered d.c. supplies.

Transistors

What is usually referred to as simply a transistor, but is more accurately described as a bipolar transistor, is a semiconductor device, which has two p–n junctions. It is capable of producing current amplification and, with an added load resistor, both a load and voltage power gain can be achieved.

Transistor

Transistor basics

A bipolar transistor consists of three separate regions or areas of doped semiconductor material and, depending on the configuration of these regions, it is possible to manufacture two basic types of device.

When the construction is such that a central n-type region is sandwiched between two p-type outer regions, a pnp transistor is formed as in Figure 8.50.

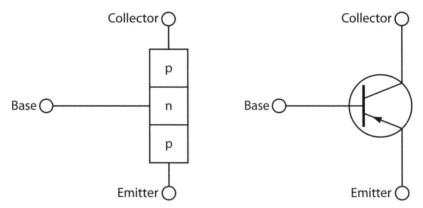

Figure 8.50   pnp transistor and its associated circuit symbols

If the regions are reversed as in Figure 8.51, an npn transistor is formed.

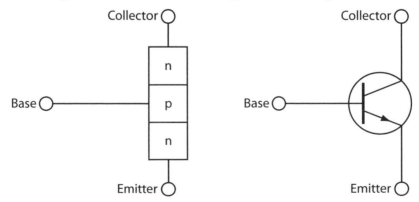

Figure 8.51   npn transistor and its associated circuit symbols

In both cases the outer regions are called the emitter and collector respectively and the central area the base.

The arrow in the circuit symbol for the pnp device points towards the base, whereas in the npn device it points away from it. The arrow indicates the direction in which conventional current would normally flow through the device.

Electron flow is of course in the opposite direction as explained in Electrical Installations Book 1. Note that in these idealised diagrams, the collector and emitter regions are shown to be the same size. This is not so in practice; the collector region is made physically larger since it normally has to dissipate the greater power during operation. Further, the base region is physically very thin, typically only a fraction of a micron (a micron is one millionth of a metre).

Hard-wire connections are made to the three regions internally; wires are then brought out through the casing to provide an external means of connection to each region. Either silicon or germanium semiconductor materials may be used in the fabrication of the transistor but silicon is preferred for reasons of temperature stability.

Transistor operation

For transistors to operate three conditions must be met:

1. The base must be very thin.

2. Majority carriers in the base must be very few.

3. The base-emitter junction must be forward biased and the base-collector junction reverse biased.

See Figure 8.52.

Figure 8.52 Transistor operation

Electrons from the emitter enter the base and diffuse through it. Due to the shape of the base most electrons reach the base-collector junction and are swept into the collector by the strong positive potential. A few electrons stay in the base long enough to meet the indigenous holes present and recombination takes place.

To maintain the forward bias on the base-emitter junction, holes enter the base from the base bias battery. It is this base current which maintains the base-emitter forward bias and therefore controls the size of the emitter current entering the base. The greater the forward bias on the base-emitter junction, the greater the number of emitter current carriers entering the base.

The collector current is always a fixed proportion of the emitter current set by the thinness of the base and the amount of doping. Holes from the emitter enter the base and diffuse through it – see Figure 8.53.

Due to the shape of the base most holes reach the base-collector junction and are swept into the collector by the strong negative potential.

Figure 8.53   Circuit diagram for operation of transistor

Current amplification

Consider, for example, as in Figure 8.54, a base bias of some 630 mV has caused a base current of 0.5 mA to flow but more importantly has initiated a collector current of 50 mA.

Figure 8.54 Current flow in transistor

Although these currents may seem insignificant, it's the comparison between the base and collector currents which is of interest. This relationship between I_B and I_C is termed the 'static value of the short-circuit forward current transfer' – we normally just call it the gain of the transistor, and it is simply a measure of how much amplification we would get. The symbol that we use for this is h_{FE}. This is the ratio between the continuous output current (collector current) and the continuous input current (base current). Thus when I_B is 0.5 mA and I_C is 50 mA the ratio is:

$$h_{FE} = \frac{I_C}{I_B}$$
$$= \frac{50 \text{ mA}}{0.5 \text{ mA}}$$

i.e. approximately equal to 100.

Note: There are no units since this is a ratio.

It can therefore be said that a small base current initiated by the controlling forward base-bias voltage produces a significantly higher value of collector current to flow dependent on the value of h_{FE} for the transistor. Thus current amplification has been achieved.

Voltage amplification

Mention was previously made to the derivation of the word 'transistor', and it quoted the device as transferring current from a low resistive circuit to approximately the same current in a high resistive circuit. This is an npn transistor so the low resistive reference is the emitter circuit and the high resistive reference, the collector circuit, the current in both being almost identical.

The reason the emitter circuit is classed as having a low resistance is because it contains the forward biased (pn) base-emitter junction. Conversely, the collector circuit contains the reverse biased (np) base-collector junction, which is of course in the order of tens of thousands of ohms (it varies with I_C). In order to produce a voltage output from the collector, a load resistor (R_B) is added to the collector circuit as indicated in the circuit diagram in Figure 8.55.

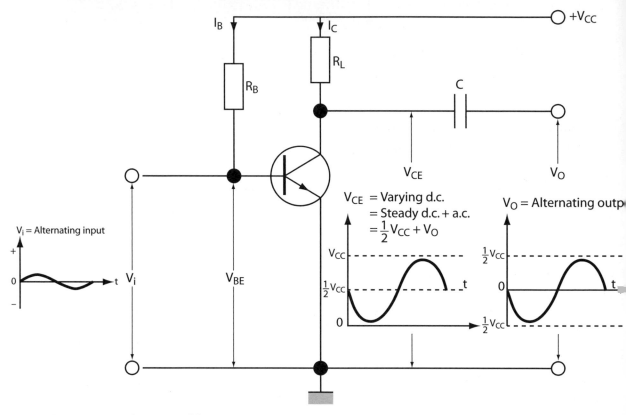

Figure 8.55 Voltage amplification

This shows the simplest circuit for a voltage amplifier. To see how voltage amplification occurs we have to consider that there is no input across V_i, which is called the quiescent (quiet) state. For transistor action to take place the base emitter junction V_{BE} must be forward biased (and has to remain so when V_i goes positive and negative due to the a.c. signal input).

By introducing resistor R_B between collector and base, a small current I_B will flow from V_{CC} through R_B into the base and down to 0V via the emitter, thus keeping the transistor running (ticking over).

Component resistor values R_B and R_L are chosen so that the steady base current I_B makes the quiescent collector-emitter voltage V_{CE} about half the power supply voltage V_{CC}. This allows V_O to replicate the input signal V_i at an amplified voltage with a 180° phase shift. When an a.c. signal is applied to the input V_i and goes positive it increases V_{BE} slightly to around 0.61 V. When V_i swings negative, V_{BE} drops slightly to 0.59 V. As a result a small alternating current is superimposed on the quiescent base current I_B, which in effect is a varying d.c. current.

The collector emitter voltage (V_{CE}) is a varying d.c. voltage, or an alternating voltage superimposed on a normal steady d.c. voltage. The capacitor C is there to block the d.c. voltage, but allow the alternating voltage to pass on to the next stage. So in summing up, a bipolar transistor will act as a voltage amplifier if:

- It has a suitable collector load R_L.

- It is biased so that so that the quiescent value V_{CE} is around half the value of V_{CC}, which is known as the class A condition.

- The transistor and load together bring about voltage amplification.

- The output is 180° out of phase with the input signal as Figure 8.55 indicates.

- The emitter is common to the input, output and power supply circuits and is usually taken as the reference point for all voltages, i.e. 0 V. It is called 'common', 'ground' or 'earth' if connected to earth.

Transistor as a switch

We have looked at the transistor as an amplifier of current and voltage. If we connect the transistor as in Figure 8.56, we can operate it as a switch. Compared with other electrically operated switches, transistors have many advantages, whether in discrete or integrated circuit (IC) form. They are small, cheap, reliable, have no moving parts and can switch millions of times per second – the perfect switch that has infinite resistance when 'off', no resistance when 'on' and changes instantaneously from one state to another, using up no power.

Figure 8.56 shows the basic circuit for an npn common emitter as in previous diagrams with a load resistor R_L connected in series with the supply (V_{CC}) and the collector.

Figure 8.56   Circuit diagram for transistor used as a switch

R_B prevents excessive base currents, which would seriously damage the transistor when forward biased. With no input across V_I, the transistor is basically turned off. This means then that there will be no current (I_C) through R_L, therefore there will be no volt drop across R_L so the +V_{CC} voltage (6 V) will appear across the output V_{CE}.

If we now connect a supply of between 2 V–6 V across V_i input, the transistor will switch on, current will flow through the collector load resistor R_L and down to common, making the output V_{CE} around 0 V. From this we can state that:

- when the input $V_i = 0$ V, the output $V_{CE} = 6$ V

- when the input $V_i = 2$ V–6 V, the output $V_{CE} = 0$ V.

From this we can see that the transistor is either High (6 V) or Low (0 V), or we can confirm, like a switch i that it is 'On' (6 V) or 'Off' (0 V).

This circuit can be used in alarms and switch relays for all types of processes and is the basic stage for programmable logic control (plc) which uses logic gates with either one or zero to represent what the output is from a possible input.

In Figure 8.57 are identified basic logic gate circuits with their inputs/outputs, 'truth table' and symbols.

Figure 8.57 Basic logic gate circuits (continued on next page)

Circuit NAND Gate

Symbol

British

American

Truth table

NAND gate
(2-input)

A	B	F
0	0	1
0	1	1
1	0	1
1	1	0

Circuit AND Gate

British

American

Symbol

British

American

Truth table

AND gate
(2-input)

A	B	C	F
0	0	1	0
0	1	1	0
1	0	1	0
1	1	0	1

Symbol

British

American

Truth table

Exclusive-OR gate

A	B	F
0	0	1
0	1	1
1	0	1
1	1	0

Figure 8.57 Basic logic gate circuits (continued)

There is also an Exclusive NOR which gives an output as indicated in Figure 8.58.

Symbol

British

American

Truth table

Exclusive NOR gate

Input		Output
A	B	
0	0	0
0	1	1
1	0	1
1	1	1

Figure 8.58 Exclusive NOR gate

Testing transistors

As all transistors consist of either a npn or a pnp construction the testing of them is similar to diodes. Special meters with three terminals for testing transistors are available and many testing instruments have this facility. However, an ohmmeter can be used for testing a transistor to check if it is conducting correctly. The following results should be obtained from a transistor assuming that the red lead of an ohmmeter is positive.

Note: This is not always the case. With some older analogue meters, the battery connections internally are the opposite way round, so it is always good to check both ways across base and emitter as shown in Figures 8.60 and 8.61.

A good npn transistor will give the following readings:

- Red to base and black to collector or emitter will give a low resistance.

- However, if the connections are reversed it will result in a high resistance reading.

- Connections of any polarity between the collector and emitter will also give a high reading.

A good pnp transistor will give the following readings:

- Black to base and red to collector or the emitter will give a low resistance reading.

- However, if the connections are reversed a high resistance reading will be observed.

- Connections of either polarity between the collector and emitter will give a high resistance reading.

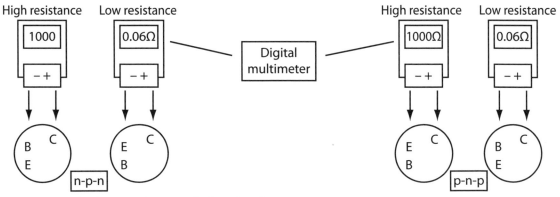

Figure 8.59 Testing transistors with digital multimeter

Figure 8.60 Testing transistors with analogue multimeter

Integrated circuits (ICs)

Integrated circuits are complete electronic circuits housed within a plastic case (known as the **black box**). The chip contains all the components required, which may include diodes, resistors, capacitors, transistors etc.

There are several categories, which include analogue, digital and memories. The basic layout is shown in Figure 8.61, which is an operational amplifier (dual in-line IC).

Figure 8.61 Operational amplifier

Definition

Dual in-line ICs – the types of IC with the pins lined up down each side

The plastic case has a notch at one end and, if you look at the back of the case with the notch at the top, Pin 1 is always the first one on the left hand side, sometimes noted with a small dot. The other pin numbers follow down the left hand side, 2, 3 and 4 then back up the right hand side from the bottom right to the top, 5, 6, 7 and 8. This is an 8-pin chip but you can get some chips with up to 32 pins or more.

Thyristors

Sometimes referred to as a 'silicon controlled rectifier' (SCR), the thyristor is a four-layer semiconducting device, with each layer consisting of an alternating n or p type material as shown in Figure 8.62.

Thyristor

Figure 8.62   Construction of a thyristor

Figure 8.63 Thyristor circuit symbol

The main terminals (the anode and cathode) are across the full four layers, while the control terminal (the gate) is attached to one of the middle layers. The circuit symbol for a thyristor is shown in Figure 8.63.

Effectively acting as a high speed switch, thyristors are available that can switch large amounts of power (as high as MW) and can therefore be seen in use within High Voltage Direct Current (HVDC) systems. These can be used to interconnect two a.c. regions of a power-distribution grid, albeit the equipment needed to convert between a.c. and d.c. can add considerable cost. That said, above a certain distance (about 35 miles for undersea cables and 500 miles for overhead cables), the lower cost of the HVDC electrical conductors can outweigh the cost of the electronics required.

Principle of operation

A thyristor acts like a semiconductor version of a mechanical switch, having two states; in other words it is either 'on' or 'off' with nothing in between. This is how they gained their name from the Greek word thyra (which means door), the inference being something that is either open or closed.

The thyristor is very similar to a diode, with the exception that it has an extra terminal (the gate) which is used to activate it. Effectively in its normal or 'forward biased' state, the thyristor acts as an open-circuit between anode and cathode, thus preventing current flow through the device. This is known as the 'forward blocking' state.

However, the thyristor can allow current to flow through it by the application of a control (gate) current to the gate terminal. It is this concept that allows a small signal at the gate to control the switching of a higher power load. In this respect the thyristor is performing in a similar way to a relay (see Book 1 Electrical science).

Once activated, a thyristor doesn't require a control (gate) current to continue operating and will therefore continue to conduct until either the supply voltage is turned off, reversed or when a minimum 'holding' current is no longer maintained between the anode and cathode.

Did you know?

The conversion from a.c. to d.c. is known as rectification and from d.c. to a.c. as inversion

These concepts are shown in the following diagram:

In Figure 8.64, switch S_1 acts as a master isolator and no supply is present at either the thyristor or at switch S_2. Closing switch S_1 will allow a supply to be present at the thyristor, but there is no signal at the gate terminal as switch S_2 is open and therefore no current will flow to the indicator lamp. However, if we now close switch S_2, the gate will be energised and the thyristor will operate, thus allowing current to flow through it to the indicator lamp. The current at the anode would be large enough in this situation to allow the thyristor to continue operating, even if we opened switch S_2.

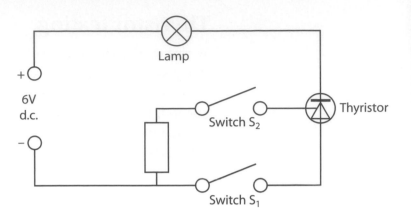

Figure 8.64   Circuit diagram of thyristor

The control of a.c. power can also be achieved with the thyristor by allowing current to be supplied to the load during part of each half cycle. If a gate pulse is applied automatically at a certain time during each positive half cycle of the input, then the thyristor will conduct during that period until it falls to zero for the negative half cycle.

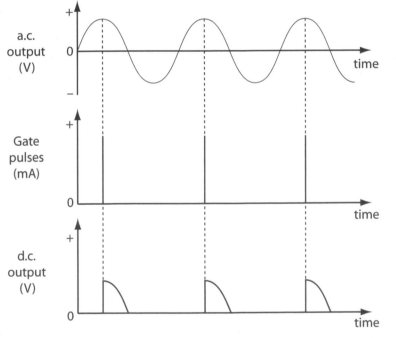

Figure 8.65   Supply of current to load during part of each half cycle

You will see from Figure 8.65 that the gate pulse (mA) occurs at the peak of the a.c. input (V). During negative half cycles the thyristor is reverse biased and will not conduct and will not conduct again until half way through the next positive half cycle. Current actually flows for only a quarter of the cycle, but by changing the timing of the gate pulses, this can be decreased further or increased. The power supplied to the load can be varied from zero to half wave rectified d.c.

Figure 8.66   Typical small value thyristor

Thyristor testing

To test thyristors a simple circuit needs to be constructed as shown in Figure 8.67. When switch B only is closed the lamp will not light, but when switch A is closed the lamp lights to full brilliance. The lamp will remain illuminated even when switch A is opened. This shows that the thyristor is operating correctly. Once a voltage has been applied to the gate the thyristor becomes forward conducting like a diode and the gate loses control.

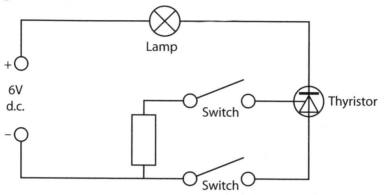

Figure 8.67   Circuit for testing thyristor

The triac

The triac is a three terminal semiconductor for controlling current in either direction and the schematic symbol for a triac is shown in Figure 8.68.

If we look at Figure 8.68 more closely, we can see that the symbol looks like two thyristors that have been connected in parallel, albeit in opposite facing directions and with only one gate terminal. We refer to this type of arrangement as an inverse parallel connection.

The main power terminals on a triac are designated as MT1 (Main Terminal 1) and MT2. When the voltage on terminal MT2 is positive with regard to MT1, if we were to apply a positive voltage to the gate terminal the left thyristor would conduct. When the voltage is reversed and a negative voltage is applied to the gate, the right thyristor would conduct.

As with the thyristor generally, a minimum holding current must be maintained in order to keep a triac conducting. A triac therefore generally operates in the same way as the thyristor, but operating in both a forward and reverse direction. It is therefore sometimes referred to as a bidirectional thyristor, in that it can conduct electricity in both directions. One disadvantage of this is that triacs can require a fairly high current pulse to turn them on.

Figure 8.68   Triac symbol

Did you know?

Triacs were originally developed for, and used extensively in, the consumer market. They are used in many low power control applications such as food mixers, electric drills and lamp dimmers, etc.

The diac

Before consideration is given to practical triac applications and circuits, it is necessary to examine the diac. This device is often used in triac triggering circuits because it, along with a resistor-capacitor network, produces an ideal pulse-style waveform. It does this without any sophisticated additional circuitry, due to its electrical characteristics. Also it provides a degree of protection against spurious triggering from electrical noise (voltage spikes).

Diac

Figure 8.69   Diac symbol

The symbol is shown in Figure 8.69. The device operates like two breakdown (Zener) diodes connected in series, back to back. It acts as an open switch until the applied voltage reaches about 32–35 volts, when it will conduct.

Lamp dimmer circuit

Figure 8.70 shows a typical GLS lamp dimmer circuit.

The GLS lamp has a tungsten filament, which allows it to operate at about 2500°C and is wired in series with the triac. The variable resistor is part of a trigger network providing a variable voltage into the gate circuit, which contains a diac connected in series. Increasing the value of the resistor increases the time taken for the capacitor to reach its charge level to pass current into the diac circuit. Reducing the resistance allows the triac to switch on faster in each half cycle. By this adjustment the light output of the lamp can be controlled from zero to full brightness.

Figure 8.70   GLS lamp dimmer circuit

The capacitor is connected in series with the variable resistor. This combination is designed to produce a variable phase shift into the gate circuit of the diac. When the p.d. across the capacitor rises, enough current flows into the diac to switch on the triac.

The diac is a triggering device having a relatively high switch on voltage (32–35 volts) and acts as an open switch until the capacitor p.d. reaches the required voltage level.

The triac is a two-directional thyristor, which is triggered on both halves of each cycle. This allows it to conduct current in either direction of the a.c. supply. Its gate is in series with the diac, allowing it to receive positive and negative pulses.

A relatively high resistive value resistor R2 (100 Ω) is placed in series with a capacitor to reduce false triggering of the triac by mains voltage interference. The capacitor is of a low value (0.1 mF). This combination is known as the **snubber circuit**.

On the job: On-call

You are asked to look at a problem at a small private residential care home. When you arrive the warden explains that there is a problem with the nurse call system, in that it has very recently been installed but, as the electrical contractor that fitted it has gone into receivership, no one is quite sure of its operation. Additionally, they require another 'patient call' button to be installed in a further bedroom. The warden has a circuit diagram of the system, which is shown below.

All components, with the exception of the patient call buttons, are located inside the nurse call panel, which is located at the nurses' station. Looking at the diagram, identify the components and prepare a written report for the warden, explaining in writing how the system operates. Now produce a revised circuit diagram to show how an additional patient could be added to the system. (Please assume that the values and ratings of any components for this exercise will be acceptable.) Your new drawing and report will then be held by the warden for future reference.

(A suggested answer can be found on page 316)

Field effect transistors (FETs)

Field effect transistor devices first appeared as separate (or discrete) transistors, but now the field effect concept is employed in the fabrication of large-scale integration arrays such as semiconductor memories, microprocessors, calculators and digital watches.

There are two types of field effect transistor: the junction gate field effect transistor, which is usually abbreviated to JUGFET, JFET or FET, and the metal oxide semiconductor field effect transistor known as the MOSFET. They differ significantly from the bipolar transistor in their characteristics, operation and construction.

The main advantages of an FET over a bipolar transistor are:

- Its operation depends on the flow of majority current carriers only. It is, therefore, often described as a unipolar transistor.

- It is simpler to fabricate and occupies less space in integrated form.

- Its input resistance is extremely high, typically above 10 MΩ especially for MOSFET devices. In practice, this is why voltage measuring devices such as oscilloscopes and digital voltmeters employ the FET in their input circuitry, so that the voltage being measured is not altered by the connection of the instrument.

- Electrical noise is the production of random minute voltages caused by the movement of current carriers through the transistor structure. Since the FET does not employ minority carriers, it therefore has the advantage of producing much lower noise levels compared with the bipolar transistor.

- Also due to its unipolar nature it is more stable during changes of temperature.

The main disadvantages of an FET over its bipolar counterpart are:

- Its very high input impedance renders it susceptible to internal damage from static electricity.

- Its voltage gain for a given bandwidth is lower. Although this may be a disadvantage at low frequencies (below 10 MHz), at high frequencies the low noise amplification that an FET achieves is highly desirable. This facet of FET operation, though, is usually only exploited in radio and TV applications, where very small high frequency signals need to be amplified.

- The FET cannot switch from its fully on to its fully off condition as fast as a bipolar transistor. It is for this reason that digital logic circuits employing MOSFET technology are slower than bipolar equivalents, although even faster switching speeds are being achieved as FET production technology continues to advance.

Remember

The main reason why the FET has such a differing characteristic from the bipolar transistor is because current flow through the device is controlled by an electric field, which is not the case with the bipolar transistor. It is for this reason that the FET is considered to be a voltage operated device rather than current operated

Figure 8.71 illustrates the basic construction of the FET, which consists of a channel of n-type semiconductor material with two connections, source (S) and drain (D). A third connection is made at the gate (G), which is made of p-type material to control the n-channel current. The symbol is shown in Figure 8.72.

Figure 8.71   Basic construction of field effect transistor (FET)

Figure 8.72   Field effect transistor (FET) symbol

In theory, the drain connection is made positive with respect to the source, and electrons are attracted towards the D terminal. If the gate is made negative there will be reverse bias between G and S, which will limit the number of electrons passing from S to D.

The gate and source are connected to a variable voltage supply, such as a potentiometer, and increasing or lowering the voltage makes G more or less negative, which in turn reduces or increases the drain current.

Component positional reference

As electronic diagrams become more complex a system called the component positional reference system is used. This system uses a simple grid reference to identify holes on a board on which components are installed. This is done by counting along the columns at the top of the board starting from the left and numbering them as you count. Then starting from the left and counting down rows, each row is given a letter in turn from the alphabet starting from A. For example the position reference point 7:J would be 7 holes from the left and 10 holes down.

Knowledge check

1. Name the four bands found in the colour code system for resistors and capacitors.

2. Give a rhyme for remembering the colours in the resistor colour code system.

3. What are the two ways that resistors can be used? Describe briefly these two ways.

4. List and describe the different types of dielectric found in capacitors and name the three factors that the value of capacitance depends on.

5. Capacitors of 10 μF, 12 μF and 40 μF are connected in series and then in parallel. Calculate the effective capacitance for each connection.

6. Describe the characteristics of a zener diode.

7. List the precautions to take with fibre optic cables.

8. Describe with the aid of diagrams the process for testing diodes.

9. Describe with the aid of diagrams voltage amplification in a transistor circuit.

10. Draw the lamp dimmer circuit and describe the operation of the individual components.

Suggested Answers

Answers to 'On the jobs'

Chapter 2 Earthing and protection

On the job: Intermittent fault page 79

The worst faults to find are intermittent ones and in an example such as this, there could be many reasons why the RCD may be tripping, most of which could never be covered by this book. However, we have stressed before that 'knowledge is power' and the customer has given us some clues that may help. Sometimes we have to think laterally to find faults, and the clues here, as a starting point for your investigation, are that the trips occur overnight and when it has been raining.

We know that there are security lights, and these will be controlled by PIR detectors, probably fitted as part of each light. The fact that the fault occurs through the night might lead us to think that one of the security lights has some kind of fault that is influenced by dampness. This fault in the light could only trip the RCD once it detects movement, which logically could only occur once it is dark. Following safe isolation procedures, an insulation-resistance test of the circuit may prove a good starting point. Also look for dead insects at terminations which may be causing the problem.

It is also possible that there is no single fault but an accumulation of problems. Immersion heaters, heating pumps and cooker elements are all known sources of small leakage (mA) to earth. Many appliances, including computer equipment, have built-in mains filtering in the form of capacitors between phase and earth and neutral and earth; computers have been recorded to leak as much as 3 mA.

Too much connected leakage current means that whereas from testing we know that a 30 mA RCD should not trip at 15 mA, a 'healthy' RCD may well trip at 16 mA if an extra problem is applied that 'pushes it over the edge'.

Another factor that is often overlooked is that a slight short between earth and neutral would not necessarily trip an RCD until a load was then applied to it. The amount of load would depend on the neutral-to-earth fault resistance. Therefore the operation of a microwave, a central-heating pump or a computer could be sufficient to trip the RCD, especially if there is already a continuous connected leakage as mentioned above.

Checking individual circuits neutral to earth can be difficult (e.g. earths all twisted together and put in one terminal) but should not be excluded from your investigation.

Chapter 3 Lighting

On the job: Fluorescent fault page 95

The discharge across the electrodes in a fluorescent fitting is extinguished about 100 times per second and this can cause a flickering effect. You normally can't see this, but when rotating machines are illuminated in this way, the stroboscopic effect can make machinery appear to slow down, change direction of rotation or even stop. You should therefore suggest the following to the factory manager. Either fit tungsten filament lamps locally to the machines, which will lessen the effect but not eliminate it completely, or, as the factory lighting will probably be split over three phases, connect adjacent fluorescent fittings to different phases of the supply. In a three-phase supply the phases are 120° out of phase with each other. Therefore, if the light falling on the machine arrives from two different sources, each will be flickering at a different time and they will interfere with one another, thus reducing the stroboscopic effect.

Chapter 4 Circuits and systems

On the job: Customer technical query page 99

You should always answer any customer enquiries politely and fully, sufficient to meet their needs. However, never mislead the customer and if you don't know the answer, say so, and promise either to find out or put them in touch with someone who does know. The basic answer would be to say that the thermostat controls the central-heating boiler by telling the programmer to switch it on when the temperature has fallen below a set level, or switch it off if it has risen above a set level.

If a more in-depth answer is required, you could say that electronic digital thermostats use a device called a thermistor to measure temperature. A thermistor is a resistor whose electrical resistance changes with temperature. The circuitry within the thermostat measures the resistance and converts it into a temperature reading.

Ideally, a room thermostat should be located in the part of the house where people spend the most time. It should be about 1.5 m above floor level and about 0.5 m away from an outside wall. It should not be exposed to any heat sources (other than the air in the room) such as sunlight, other appliances, heater vents, windows or hot-water pipes. Consequently, it is best located in a living room rather than the entrance hall (as is very commonly done), as the hall temperature can be lowered by the front door being used and the heating may, therefore, be running more often than is actually required.

Chapter 5 Inspection, testing and commissioning

On the job: Periodic inspection page 183

When you arrived at the premises, you should have fully explained the purpose and implications of your visit to someone in charge. Ideally, all aspects of your work should have been explained to the customer even before you arrived.

Whenever diagrams, charts or tables are not available, some exploratory work should be undertaken to make sure that you could carry out the inspection and testing safely and effectively. Check that alterations such as the extension have not affected the suitability of any of the wiring for its present load and method of installation. Remember that with periodic inspection and testing, inspection is the vital initial operation and that testing subsequently carried out is in support of that inspection. For safety it is necessary to carry out a visual inspection of the installation before beginning any tests or opening enclosures, removing covers etc.

Whenever electronic equipment such as a computer is in use, arrangements should also be made with the customer to have it disconnected for the period of the test. You should therefore try to make this arrangement now. However, remember that some IT systems use back-up power supplies, so check isolation carefully. Carry out your work and complete a Periodic Inspection & Test Certificate with the Schedule of Results sheet attached.

Chapter 6 Fault diagnosis and rectification

On the job: Fault diagnosis page 226

The secret of fault finding is to look logically for the differences between what is there and what should be there. So stay calm, don't panic and you are half way there! In a fictitious example such as this we can never be sure what the fault is, but let's start by reasoning out what should happen.

We know that the motor is only trying to start from the remote station. We also know that the motor should run when we take our finger off the start button at the remote station. The fact that it starts at all from the remote station indicates that a supply is present at the start button, but what does that start button actually do? Well, normally the start button in a remote station is a 'normally open' contact connected in parallel across the start button in the starter/isolator. The feed from the DB normally goes straight to the remote station start button, passing firstly through the 'normally closed' stop button inside the remote station. As the two start buttons are connected in parallel, one of the wires should create a permanent supply at the start button in the starter/isolator. When the remote start is pressed, normally the supply passes from one side of the button to the other, where that parallel wire goes to the other side of the start button in the starter/isolator to energise the coil. This energises the coil and the 'hold in' contact comes into play, thus maintaining the permanent supply to the coil when either start button is released.

We should therefore start our investigation by looking to see whether the parallel wire taking the permanent supply between start buttons has been disconnected or is broken.

Chapter 7 Motors, generators and motor control

On the job: Motor noise page 259

Noise indicates motor problems, but in itself does not cause damage. Noise, however, is usually accompanied by vibration, and vibration can cause damage in several ways. It can shake windings loose, damage insulation mechanically by cracking the material, cause brush sparking at commutators or current collector rings and speed up bearing failure.

Consequently, whenever noise or vibration is reported in an operating motor, the source should quickly be investigated and corrected. Noise and vibrations can be caused by a misaligned motor shaft or can be transmitted to the motor from the driven machine or power transmission system. They can also be the result of electrical imbalance in the motor.

So, having checked the motor shaft alignment, disconnect the motor from the driven load. If the motor then operates smoothly, look for the source of noise or vibration in the driven equipment. If the disconnected motor still vibrates, remove the supply from the motor. If the vibration stops, look for an electrical imbalance. If it continues as the motor coasts without power, look for a mechanical imbalance.

Electrical imbalance occurs when the magnetic attraction between stator and rotor is uneven around the periphery of the motor. This causes the shaft to deflect as it rotates, thus creating a mechanical imbalance. Electrical imbalance usually indicates an electrical failure such as an open stator or rotor winding, an open bar or ring in squirrel-cage motors or shorted field coils in synchronous motors. An uneven air gap, usually resulting from badly worn sleeve bearings, also produces electrical imbalance.

Chapter 8 Electronics

On the job: On-call page 310

The components used are five push buttons (N/O & N/C), three diodes, a thyristor, a resistor, three indicator lamps and a relay to control the buzzer.

Any patient pressing their button will send a signal to the gate of the thyristor which, as it has a permanent supply to it, will allow the supply to energise the coil of the relay, in turn activating the buzzer. The same signal from each push button will activate the respective indicator lamp to show which patient is calling, and the diodes prevent any signal 'backfeeding' onto any of the other buttons, thus preventing all of the indicators from being lit. The resistor limits the current flow in case more than one patient activates the system at the same time.

Push-button B then acts as an isolator to silence the sounder, but not the indicators, and button A isolates the entire system. Each individual patient's call button will be reset by the nurse once they have attended to the situation.

Glossary

Words in bold mean there is also a glossary entry for that term. *See also* refers to a closely related term.

a.c.   Alternating current, but often used to describe anything that uses it, e.g. a.c. voltage, a.c. motor (*See also* **d.c., sine wave**)

active or true power   **Power** consumed by a resistor, dissipated as heat and not returned to the source (*See also* **apparent power, reactive power**)

adiabatic equation   Used to check that a **conductor** will carry a **fault current** without overheating

adjacent   In trigonometry, the side of a triangle adjacent to the angle under consideration and the right angle

ambient temperature   Surrounding temperature within the area where an electric cable is installed

ammeter   Instrument for measuring **current**; can be **analogue** or **digital**

analogue device   Represents measurement by position of, for example, a needle on a calibrated scale (*See also* **digital device**)

apparent power   Sum of **true power** and **reactive power** in an **a.c.** circuit

armature   Moving part of electrical machine in which a **voltage** is induced by a magnetic field (*See also* **rotor, stator**)

asynchronous   Used to describe motors in which the speed of the motor and of the rotating magnetic field are different (*See also* **synchronous, slip**)

average value   For **sinewave a.c.** this equals either **peak value** or **voltage** × 0.637 (*See also* **r.m.s.**)

ballast (or choke)   **Inductance** coil in fluorescent lights to limit variations of **a.c.** or alter its **phase**

black box   Plastic case housing complete electronic circuit (*See also* **integrated circuit**)

capacitance   Ability of circuit or device to store electric **charge**, measured in farads – generally microfarads (μF), nanofarads (nF) or picofarads (pF); symbol C (*See also* **capacitor**)

capacitive   Circuit containing components that can store electric charge, i.e. have capacitance

capacitive reactance   Opposition to flow of a.c. produced by a **capacitor**, measured in ohms (Ω); symbol X_C

capacitor   Component with ability to store required quantity of electric **charge** if **voltage** applied to it, which can be returned to a circuit (*See also* **capacitance, dielectric**)

CCD   Camera system used in **CCTV** system, more expensive than **CMOS** but images clear and sharp

CCTV   Closed circuit television, used in security systems

charge   Quantity of electrons, e.g. as stored in a **capacitor**, measured in **coulombs**; symbol Q

choke (or ballast)   **Inductance** coil in fluorescent lights to limit variations of **a.c.** or alter its **phase**

coercivity   Due to **hysteresis**, reverse magnetic **field strength** that must be applied to reduce **flux density** to zero (*See also* **remanence**)

CMOS   Camera system used in **CCTV** system, cheap but images not clear or sharp (*See also* **CCD**)

commutator   Device for connecting a rotating **current** carrying coil to a **d.c.** supply

conductor   Any material allowing electric charges to flow freely, including metal pipes, metal structures of buildings, salt water and ionised gases (*See also* **semiconductor, insulator**)

contactor   Widely used in electrical installations, using **solenoid** effect to make or break contacts

control panel   Programmable 'brains' of a system to which all parts are connected

cosine   In a right-angle triangle, ratio of length of side **adjacent** to an angle to length of the longest side (**hypotenuse**), abbreviated to 'cos' (*See also* **power factor**)

coulomb   SI unit of electric charge, the quantity of electricity transported in one second by **current** of one ampere; symbol C

cpc   Circuit protective conductor

csa   Cross-sectional area, frequently used when describing **conductors**

current   Flow of free electrons in a **conductor**, measured in amperes (A); symbol I; direction is opposite to actual electron flow

d.c.   Direct current, but often used to describe anything that uses one, e.g. d.c. voltage, d.c. motor (*See also* **a.c.**)

delta connection   Triangular arrangement of electrical three-phase windings

dielectric   Type of **insulator**, especially used as thin film in **capacitors**, in which charges are displaced when subjected to a **potential difference** but do not flow

digital device   Represents measurement as numbers (liquid crystal or **LED** display) rather than a needle on a scale as in **analogue device**

diode   **Semiconductor** device that allows current to flow in one direction only

directly proportional   If two properties are directly proportional then an increase in one will cause a corresponding increase in the other (*See also* **indirectly proportional**)

discrimination   Correct arrangement of fuses in sequence to protect parts of a circuit

DOL   Direct-on-line, type of motor starter arrangement

dual-in-line IC   Integrated circuit with pins in line down each side

EEBADOS   Earthed equipotential bonding and automatic disconnection of the supply

e.m.f.   *See* **electromotive force**

electromotive force   Greatest potential difference that can be generated by a current source, measured in volts (*See also* **potential difference** and **voltage**)

fault current   Current due to a fault in a circuit, including earth-fault current

FET   Field-effect transistor, type of transistor common in integrated circuits

field strength   Force field around a magnet or current carrying **conductor**, measured in teslas (T); symbol H (*See also* **flux density**)

Fleming's left-hand (motor) rule   If the first finger, the second finger and the thumb of the left hand are held at right angles to each other, then the first finger pointing in the direction of the Field (N to S) and the second finger pointing in the direction of the current in the conductor, the thumb will now be indicating the direction in which the conductor tends to move

flux density   Quantity of magnetism taking account of magnetic **field strength** and its extent, measured in webers; symbol B

former   Block of material, normally cylindrical or cuboid, around which wire is wound in a helical fashion to produce coils or windings

frequency (f)   Rate at which **a.c.** goes from zero to positive maximum, to negative maximum and back to zero (one cycle), measured in hertz (Hz)

fused spur   Spur connected to ring final circuit through fused connection

fusing factor   Ratio of minimum current causing a fuse or circuit breaker to trip and stated current that either can sustain without blowing

HBC   High breaking capacity, BS 88 sophisticated variant of a cartridge fuse

HSE Health and Safety Executive

hypotenuse Side in a right-angle triangle opposite to the right angle (*See also* **sine**, **cosine**, **Pythagoras' theorem**)

hysteresis Lagging of an effect behind its cause, e.g. change in magnetism of a body which lags behind changes in magnetic field (*See also* **coercivity**, **remanence**)

IC *See* **integrated circuit**

illuminance Measure of visible light reaching a surface, measured in lumens per square metre (lux); symbol lx

impedance Total opposition to flow of **a.c.** in a circuit, combining **resistance**, **inductance** and **capacitance**, measured in ohms (Ω); symbol Z

indirectly proportional If two properties are indirectly proportional then an increase in one will cause a corresponding decrease in the other (*See also* **directly proportional**)

inductance Opposition created by a changing **current** in a magnetic field which induces a **voltage** to oppose change in current, either within a circuit (self-inductance) or a neighbouring circuit in the same magnetic field (mutual inductance), measured in henrys (H); symbol L (*See also* **mutual inductance**)

inductive circuit or **load** Containing components with windings, e.g. motor, generator or transformer, which have **inductance**

inductive reactance Opposition to flow of **a.c.** produced by an inductor, measured in ohms (Ω); symbol X_L

inductor Component introduced into a circuit to provide required amount of inductance

infrared Electromagnetic radiation below the wavelength of visible light that we sense as heat, used in sensors for intruder alarm systems (*See also* **PIR**)

in phase Where **current** and **voltage** alternate at the same time (*See also* **out of phase**)

insulator Poor conductor of electric charges (i.e. a **current**) or heat (*See also* **conductor**, **dielectric**, **semiconductor**)

integrated circuit Miniature electronic circuit produced on a single **semiconductor** crystal offering low cost, bulk, reliability and high speed

Invar Nickel-iron alloy with lowest known rate of expansion from room temperature to 230°C, widely used in instruments, oven thermostats etc.

isolator Mechanical switch to separate an installation from all sources of electrical energy

IT Information technology, generally computers and computer equipment

JIB Joint Industry Board for the Electrical Contracting Industry: sets procedures, particularly for safety

LED Light emitting **diode**: **semiconductor** device that emits light when **current** flows through it, often used in instrument displays

line current **Current** flowing in any one **phase** of a three-phase circuit between source and load

line voltage **Potential difference** between any two-phase conductors between source and load in a three-phase electrical circuit

MCB *See* **miniature circuit breaker**

MCCB Moulded case circuit breaker, protective device usually integral to switchgear

M$_i$ Mineral-insulated cable

miniature circuit breaker automatic switch which opens when excess **current** flows and can be closed again when current returns to normal

multimeter Single instrument (**analogue** or **digital**) combining **ammeter**, **voltmeter**, **ohmmeter**, **wattmeter** etc.

mutual inductance Production of **e.m.f.** in a circuit by a change in **current** in an adjacent circuit (*See also* **inductance**)

no volts protection Prevention of automatic restarting should a motor stop because of supply failure

non-fused spur Usually directly connected to a circuit at the terminal of socket outlets (*See also* **spur**, **fused spur**)

ohmmeter Instrument for measuring **resistance**, can be **analogue** or **digital**

Ohm's Law **Current** flowing in a circuit is directly proportional to the **voltage** applied and indirectly proportional to circuit **resistance**, provided temperature remains constant (I=V/R or V=IR or R=V/I)

oil dashpot Used in **MCB** construction to achieve time-delay tripping

out of phase Where **current** and **voltage** are alternating at different times (*See also* **in phase**)

overcurrent Current exceeding rated value or current-carrying capacity of a **conductor**

overload current Overcurrent occurring in a circuit that is electrically sound

p.d. *See* **potential difference**

panel *See* **control panel**

peak value Maximum value of **a.c.** or **d.c.** wave form, positive or negative, value from maximum to minimum is known as 'peak to peak'

PELV Protective extra-low voltage system, electrically separate from earth and other systems to protect against direct and indirect contact (also **SELV**)

PEN Protective earth neutral

phase *See* **in phase** and **out of phase**

phase angle Difference in **phase** between two sinusoidally varying quantities (*See also* **power factor**); symbol Ø

phase current **Current** through any one component comprising a three-phase source or load

phase voltage **Voltage** measured across a single component in a three-phase source or load

phasor Straight line with length representing size of an **a.c.** quantity and direction representing relationship between **voltage** and **current** (*See also* **phase angle**)

PIR Passive infra red, commonly found as PIR sensor in alarm systems (*See also* **infra red**)

PME Protective multiple earth, the most commonly used earthing system (others are **TN-S**, **TN-C-S**, **TT**)

pole pair A magnet always comprises a north and south pole; hence we always calculate using 'pole pairs'

potential difference Difference in electrical **charge** between two points in a circuit; or force available to push **current** around a circuit; measured in volts, symbol V (*See also* **voltage**, **e.m.f.**)

power Energy used per second, measured in watts (W); symbol P (*See also* **power factor** and **reactive power**)

power factor Number less than 1.0, ratio of apparent to true power in a circuit; equals cos Ø (*See also* **cosine**, **phase angle**)

Pythagoras' theorem In a right-angle triangle, if you square the lengths of the two short sides and add them, the sum equals the square of the length of the long side (**hypotenuse**)

reactive current Current in a **capacitive** or **inductive circuit**, with no resistance and no dissipation of energy

reactive power Power 'consumed' by a **capacitor** or **inductor** is not dissipated as heat but returned to source when **current** reverses (*See also* **apparent power**, **true or active power**)

remanence Due to **hysteresis**, when an applied magnetic **field strength** reduces to zero some **flux density** remains (*See also* **coercivity**)

r.m.s. Root mean square value is the effective **voltage** or **current**, defined as the **a.c.** equivalent of a **d.c.** quantity delivering the same average power to the same resistor, equals 0.707 X peak current or voltage (*See also* **peak value**, **average value**)

RCBO Residual current device with overload protection, which combines the capabilities of an **RCD** and an **MCB**

RCD *See* **residual current device**

reactance *See* **capacitive reactance**, **inductive reactance**

reactive power Power returned to source by reactive components in a circuit (**capacitors**, **inductors**), measured in volt-amperes-reactive; symbol V_{AR}

remanence **Flux density** remaining in a magnetic material after the magnetic field has reduced to zero (*See also* **hysteresis**)

residual current device Monitors **current** flowing in phase and neutral **conductors** and trips if an earth-fault creates a difference above a predetermined value (*See also* **fault current**)

resistance Opposition to flow of electrons (**current**) in a circuit, measured in amperes (A); symbol R

resistive circuit Circuit containing no **capacitance** or **inductance** (in practice there is always some), whereby voltage and current are in phase

resistor Component in a circuit to give a desired amount of **resistance**

rotor Part of a motor or generator that rotates in a magnetic field; it may carry conductors or the magnetic field system (*See also* **stator**, **armature**)

SELV Separate extra-low voltage system, electrically separate from earth and other systems to protect against direct and indirect contact (also **PELV**)

semiconductor Material (commonly silicon, germanium) that will conduct under certain conditions, allowing electrons (negative charge) or 'holes' (positive charge) to flow (*See also* **diode**, **transistor semiconductor**)

short-circuit current Overcurrent resulting from a fault that creates negligible **impedance** between live conductors (*See also* **overcurrent**)

simmerstat Energy regulator to control temperature of electric cooker plates using a bi-metal strip

sine In a right-angle triangle, the ratio of the length of the side opposite to an angle to the length of the longest side (**hypotenuse**), abbreviated 'sin' (*See also* **sine wave**)

sine wave Path traced on a graph by a pure **a.c. current** or **voltage** going from zero to positive maximum, to negative maximum and back to zero (one cycle) (*See also* **sine**, **frequency**)

slip In a motor or generator, the difference between the speed of the rotor and the speed of the rotating magnetic field, measured as a fraction (s) or percentage (S)

solenoid Coil of wire on a cylindrical former which, when **current** is passed through it, creates a magnetic field parallel to the axis to operate a plunger or other device (*See also* **contactor**)

spur Radial branch from a ring final circuit, perhaps to feed a socket outlet

star connection Y-shaped arrangement of three-phase electrical windings (*See also* **delta connection**)

star-delta System sometimes used in induction motors, whereby a **star connection** switches to a **delta connection**

static Electric charge caused by an excess of electrons (negative charge) or a deficiency (positive charge) on a conducting or non-conducting surface, creating a **potential difference** (volts)

stator   Stationary part of a motor or generator which rotates in a magnetic field; it may carry **conductors** or the magnetic field system (*See also* **stator**, **armature**)

stroboscopic   Effect when the flicker rate in a light coincides with the rotation of an object and causes it to look as if it has slowed down or stopped

synchronous   Used to describe motors in which the speed of the motor and of the rotating magnetic field are the same (*See also* **asynchronous**, **slip**)

tan   Abbreviation for tangent, in a right-angle triangle the ratio of the length of the side opposite an angle to the length of the **adjacent** side

thermistor   Device in which **resistance** quickly increases (PTC type) or decreases (NTC type) as temperature rises

thermostat   Temperature controller for ovens, central heating etc.

thyristor   **Semiconductor** device enabling high-speed switching, formerly known as a silicon-controlled rectifier (SCR)

TN-C-S system   Type of earthing system (others are **TT**, **TN- S**, **PME**)

TN-S system   Type of earthing system (others are **TT**, **TN-C-S**, **PME**)

torque   Turning force, such as that produced by an electric motor on its rotating shaft; measured as force X distance from its point of rotation

transformer   Transfers electrical energy from one **a.c.** circuit to another by **mutual inductance** between two stationary coils wound onto a former; output to input values (e.g. voltage) are determined by the relative number of turns in each coil

transistor semiconductor   Device that allows a larger **current** to be controlled by a smaller one, allowing **voltage** or **current** amplification or as a switch

trimmer capacitor   Flat metal leaves separated by plastic film, screwed inwards to vary **capacitance** by a small amount (*See also* **tuning capacitor**)

true or active power   Power consumed by a resistor, dissipated as heat and not returned to the source (*See also* **apparent power**, **reactive power**)

TT system   Type of earthing system (others are **TN-S**, **TN-C-S**, **PME**)

tuning capacitor   Interleaved metal plates, one set fixed, one movable; **capacitance** controlled by area overlapping (*See also* **trimming capacitor**)

UPS   Uninterruptible power supply, used especially with computer systems

voltage   Difference in electrical **charge** between two points in a circuit; or force available to push **current** around a circuit; measured in volts, symbol V (*See also* **potential difference** and **electromotive force**)

voltage drop   Loss of **voltage** (electrical pressure) due to **resistance** in **conductor** and components in a circuit

voltmeter   Instrument for measuring voltage

wattmeter   Instrument for measuring power

windage   Physical resistance experienced by rotating parts of a motor or generator created by air within the casing

Index

A

a.c. circuits 19–25
  capacitive 10, 14, 16, 20, 22
  capacitor 8–9
  impedance 10–13
  inductance 6–8
  inductive 10, 11, 22
  phase angle 5, 18, 21
  power calculations 23–5, 39–42
  power factor 5, 7–8, 21, 22
  power triangle 22–3
  resistance 5–6
  resistance and capacitance in series (RC) 14–15
  resistance, inductance and capacitance in parallel 18–19
  resistance, inductance and capacitance in series (RLC) 16–17
  resistance and inductance in series (RL) 11–12
  resistive 10, 19–21
  three-phase power 23–5, 39–40
  see also alternating current
a.c. drive 259
a.c. generator 232
a.c. motors 233–43, 246
  armature 234
  capacitor start-capacitor run motor 242–3
  capacitor-start motor 241–2
  permanent split capacitor (PSC) 242
  series-wound (universal) 233–4
  shaded pole motor 243
  single-phase induction motor 238, 245
  speed 246–8, 259
  split-phase motor 239–40
  squirrel cage 236–7, 244, 245, 255
  starters 251–7
  stator 232, 233, 235, 236, 246
  synchronous induction motors 249
  three-phase induction motors 234–8, 244
  windings 250
access requirements 166, 220–1
accessories and equipment 163, 204
adiabatic equation 79–81
alarm systems see fire alarm; intruder alarm
alternating current theory 2–25
alternating waveform
  frequency 4–5
  peak value 3, 4
  periodic time 5
  r.m.s. or effective value 3–4
  values of 2–5
ambient temperature 146, 166
  and cable selection 69, 71–2
ammeter 21, 35, 36, 41
antistatic precautions 222
apparent power 7, 10, 21, 22, 41
arc tube 95
Assumed Current Demand 83
auto-transformer starter 257
automatic star-delta starter 256
autotransformer 29

B

black box 305
British Standards 137, 142, 153
  emergency lighting 101, 102
  fire alarm systems 103, 108, 109
  initial inspection 162, 163, 164
  inspection/testing 122, 123–4, 127, 130, 131, 171, 173
  lightning protection 52
  periodic testing 152, 180–1
  sequence of tests 132
  test instruments 129, 148, 149–50
BS 88 high breaking capacity (HBC) fuses 60
BS 1361/1362 cartridge fuses 59–60
BS 3036 fuses 58–9, 71, 73
  problems with 65

C

cable routes inspection requirements 163
cable selection 69–84, 108, 163
  adiabatic equation 79–81
  ambient temperature 69, 71–2
  calculations 78, 81–2
  correction factors 71–2
  design current 70
  diversity 82–3
  external influences 69–70
  grouping factors 71, 72
  mineral insulated 71, 73, 108, 203
  rating of protective device 70–1
  shock protection 76–8
  thermal constraints 79–82
  thermal insulation 71, 72
cables
  inspection of 122, 175, 203–4
  interconnections 200–1
  joints 201
  seals and entries 203–4
  terminating devices 201, 202
calibration of instruments 148–9
cameras, CCTV 116
capacitance 8–9, 20, 274–5
capacitive circuits 10, 14, 20, 22
  risk of high frequency 224
capacitive reactance 8, 10
capacitor smoothing 294
capacitor start-capacitor run motor 242–3
capacitor-start motor 241–2
capacitors 21, 92, 206, 274–84
  basic principles 274–5
  calculations with 281
  charging and discharging 283–4
  coding 277–80
  in combination 281–2
  electrolytic 275
  fixed 275–7
  non-electrolytic 276
  polarity 280
  tantalum 275, 280
  trimmer 277
  tuning 277
  variable 277
  working voltage 274, 277
caravan installations 178
carbon brushes 234
cartridge fuses 59–60
cathode shields 90
cathodes 90
ceiling rose 89, 207
certification 128–30, 152–60
choke 90, 91, 92
choke smoothing 295
Cinematography (Safety) Regulations 173
circuit boards 206
circuit breakers 57
  breaking capacities 63
  inspection requirements 176
  MCBs 61–3
  operation of 64–5

testing 178, 180, 183
see also overcurrent protection
circuit protective conductor (cpc) 46, 79–80
  continuity of 128, 133–4, 135–6
  testing 170, 178, 184
circuits
  Band 1 and 2 165
  installation of several 193–4, 195
  see also a.c. circuits
cistern-type water heater 111
closed-circuit television 115–17
  light levels 116
  monitoring and recording 117
  wireless 115–16
coercivity 28
commissioning 127, 130–1
  handover to client 131
commutator 230–1
component failure 198
component positional reference 312
compound motor 231, 258
compression joints 201
computer disconnection 174, 224
conductors
  equipotential 178
  inspection of 122, 158, 162–3, 175, 203
  and magnetic fields 228–9
  ring circuit 178
conduits: inspection checklists 169–70
connections: inspection
  requirements 175
connectors 201
Construction (Design and Management)
  Regulations 130, 156, 171
contactor starter 251–3
contactors: faults 205
continuity testing 183, 184, 192
  cpcs 128, 133–4
  equipotential bonding 213
  protective conductors 178, 179, 184, 213
  ring circuit 135–6, 158, 178, 185, 186
  ring final circuit conductors 213, 214
control equipment faults 204–5
control panels 105, 106, 118, 205
cooker thermostats/controllers 113–15
core-type transformers 27–8
correction factors 71–4, 146
  application of 73–4
current amplification: transistors 299
current flow and magnetic fields 228–9
current, measuring 35
current transformers 30
customers and repair issues 217

D

damage due to over voltage 223
Danfoss Randall Control Packs 98
d.c. circuits 39, 295
d.c. generator 232
d.c. motor 230–1
  armature 230–1
  commutator 230–1
  reversing 231
  speed control 258–9
  types of 231
diac 309
diagrams 166, 212
  see also phasor
dimmer controls 89
diode 285, 286–90, 292, 293
direct contact protection 124, 156, 176
  inspection requirements 163–4, 165
Direct-On-Line (DOL) starter 251–3
discharge lighting 90–5
  arc tube 95
  colour appearance 91
  colour rendering 91
  high-pressure mercury vapour 95
  high-pressure sodium 95
  lamp efficacy 91
  low pressure mercury vapour 90–4, 95
  low-pressure sodium 95
disconnection hazards 174
discrimination 68, 207
documentation, inspection of 125
dual in-line ICs 305

E

earth conductor 46, 79–80
earth faults 162, 196, 206
earth-fault loop impedance 46, 54
  British Standard Tables 144
  and cable selection 76–7
  and overcurrent devices 144–6
  standard PVC circuits 144
  testing 128, 142–6, 160, 178, 179, 183, 215–16
  unsatisfactory test results 185
  verifying test results 144
earth-fault loop path 44, 46, 56–7
earthed equipotential bonding and
  automatic disconnection (EEBAD) 53, 54, 162, 176
earthing 43, 47–51, 164
  earth electrode resistance
    testing 140–2
  earth-loop impedance testing 150, 158

good and bad earth path 45
IT system 51
on the job answer 314
labelling 177
lightning protection 52
Protective Multiple Earthing 49, 50–1
purpose of 43, 44–6
results of unearthed appliance 44
and supply characteristics 182
TN-C system 51
TN-C-S system 49–50
TN-S system 48–9
TT system 47–8
electric shock 44, 45, 52–3
  and cable selection 76–8
  direct contact 52–3
  indirect contact 53
  protection 163–4
Electrical Contractors Association
  (ECA) 152
Electrical Installation Certificate 128, 130, 153, 160
  Guidance Notes 155–6
  (Type 1) 152, 154
  (Type 2) 152, 155–6
Electricity at Work Regulations (1989) 121, 125, 130, 171, 172, 173, 174
Electricity Supply Regulations (1988) 130, 173
electro-static discharge 222–3
electromagnetic force 228–9
electronics 261–313
  diagrams 312
  faults in equipment 206
  on-call answer 316–17
  protection of equipment 223–4
  see also capacitors; resistors;
    semiconductor devices; transistors
emergency lighting 100–3, 173
  battery back-up 101
  duration 101
  illuminance levels 102
  maintained/non-maintained 101
  servicing of 103
  siting of luminaires 101–2
  standby lighting 102
  sustained 101
emergency-escape lighting 100, 102
enclosures and mechanical
  protection 176
environment 166
  and cable selection 69–70
equipment 198
  external influences 166

equipotential bonding conductors 178, 179
erection methods, inspection 166
escape-route lighting 100, 101
EU Standards: emergency lighting 102
exposed conductive parts 54, 110
extraneous conductive parts 54, 110

F

Faraday's Law 237
fault current 58
fault detector coil 54, 55
fault diagnosis 187–226
  access to system 220–1
  antistatic precautions 222
  damage due to over voltage 223
  electro-static discharge 222–3
  fibre-optic cable 222
  functional testing 215–16
  hazards 222–3, 224
  interpretation of results 213–15
  isolation 188–9, 221, 225
  on the job answer 315–16
  provision of temporary supply 221
  restoration of supply 192
  special requirements 220–1
  stages of 211–12
fault protection 193–4, 195
faults
  cable interconnections 200–1
  categories of 193–9
  complete loss of supply 195
  electrical system 209
  information required 209, 210, 212–13
  installer misuse 198
  insulation failure 197
  localised loss of supply 195
  misuse, abuse and negligence 198
  plant, equipment and
    component 198
  position of 193–5
  prevention by maintenance 199
  short circuit 194, 196
  stages of diagnosis and
    rectification 211–12
  transient voltages 197
  types of 200–8
  use of other personnel 210
  see also repair process
fibre optic cable 222, 291–2
field effect transistors (FETs) 311–12
filter circuits 295
fire alarm systems 103–9
  addressable 105
  alarm sounders 107–8

automatic detectors 103, 106–8
break-glass contacts 103, 104, 106
cable selection 108
classification of 103, 104
conventional 104–5
fire compartment 106
heat and smoke detectors 107
life protection 104
maintenance checks 109
property protection 103–4
radio addressable 105
regulations 173
standby back-up 108
wiring systems 108
zones 105–6
fire prevention 105, 163, 176
Fleming's left-hand rule 229
flexible cables and cords 175
  faults 207, 208
fluorescent lamps 90–4, 95
  faults 198
  high frequency circuits 93, 94
  on the job answer 314–15
  quick-start 94, 207
  starters 91–2, 94
  stroboscopic effect 93–4
  thermal starter circuit 94
  see also discharge lighting
FP200 cable 203
fused plugs 192, 207
fuses 57, 58–60, 64–5, 71, 73, 205

G

general series circuit 16–17
generators 232
GES lamp cap 88
glow-type starter circuit 91–2
GLS lamp 86–7
  dimmer circuit 309
  lamp caps 88
grouping factors 71, 72
Guidance Note GS 38 190

H

hand-operated star-delta starter 254–5
Health and Safety at Work Act (1974) 125, 130
Health and Safety Executive 171–2, 190
heating systems 98–9
Heatplan Packs (HPP) 98
Heatshare Packs (HSP) 98
high frequency circuits 93, 94
High Voltage Direct Current Systems 306
high-risk-task area lighting 100, 101, 102–3

hot water systems 98–9
hysteresis 28

I

IEE Guidance Notes 3 152, 161, 174
  inspection frequency 172
IEE On-Site Guide 82–3, 142, 152, 161
  earth-fault loop impedance 144, 145–6
  lighting circuits 89
IEE Wiring Regulations 156, 181, 200
  cable selection 70–1, 72, 74, 75
  functional testing 215
  inspection and testing 127, 161
  insulation resistance 214
  live and non-live tests 213
immersion heaters 110, 207
impedance (Z) 10–13
incandescent lamps 86–8
indirect contact protection 124, 176
  inspection requirements 164–5
inductance 6–8
inductive circuit 10, 11–12, 22
inductive reactance 6, 10, 12–13
inductor 10, 11
inertia switch 118
infrared heaters 112
infrared source and sensor 290
initial inspection 161–6
  scope 162
initial verification 127–8
inspection checklists 168–70
Inspection Schedule 152, 156–7, 167
inspection and testing 121–86
  alterations and additions 128
  anticipation of danger 175
  continuity of conductors 128, 133–4, 135–6
  earth electrode resistance 140–2
  earth-fault loop impedance 142–6
  frequency of 122
  information required 129–30, 166
  initial inspection 161–6
  initial verification 127–8
  instruments 129, 185
  insulation resistance tests 137–9
  items covered 122–5, 126
  periodic 125–6, 171–83
  polarity 140, 157, 160, 163
  random samples 126, 175
  recording results 128, 148
  routine checks 172–3
  safety issues 129, 148
  Schedule of Test Results 128, 152, 153, 155, 157–8
  sequence of initial tests 132–51

unsatisfactory results 184–5
visual inspection 125, 161
instantaneous water heater 111
instrument transformers 29
instruments 35–41, 148–9
  all in one 151
  ammeter 35, 36, 41
  analogue and digital 38, 148
  battery checking 148
  calibration and accuracy 148–9, 190
  checklist for 216
  faults 206
  limitation and range 216
  measuring power factor 41
  multimeter 35, 36, 304
  ohmmeter 35, 36
  RCD testers 151
  test 129, 148, 190
  using meters 38
  voltmeter 35, 36, 37, 41
  wattmeter 21, 39, 40, 41
Insulated Gate Bipolar Transistors
  (IGBT) 259
insulation failure 197
insulation resistance 150, 158
  tests 137–9, 160, 178, 180, 183, 213,
    214–15
  unsatisfactory test results 184–5
integrated circuits (ICs) 305
intelligent systems 105
intruder alarm systems 118–19, 290
  audible and visual warning
    systems 118
Invar rod 115
isolating and switching devices 165
  testing 178, 183
isolation
  complete installation 188
  devices 165, 191
  fault diagnosis 188–9, 225
  individual circuit/equipment 189
  locking off device 191
isolators 180, 192, 215
IT earthing system 51
IT systems 224

J

joint boxes checklists 168
Joint Industry Board (JIB) 188
joints and connections 175
  cable joints 201
junction boxes 201

K

kilovolt-amps (kVA) 31

L

labelling 148–9, 166, 177, 183
lamp caps 88
lamp efficacy 91
lead acid cells hazard 224
legal responsibility: repairs 219
light emitting diodes (LEDs) 288
lighting 85–96
  discharge 90–5
  emergency 100–3
  faults 198
  GLS lamp dimmer circuit 309
  incandescent lamps 86–8
lighting controls checklists 169
lighting points inspection 169, 175
lightning strikes 52, 197
linear tungsten halogen lamps 87–8
live conductor 66
local authority licence 173
luminaires: faults 207

M

magnetic fields 228–9
main switches, testing 180
maintenance 193, 199
marking and labelling 177
Maximum Demand and Diversity 82–4
measurements
  current 35
  and instruments 35–41
  loading errors 37
  power 39–42
  resistance 35, 36
  voltage 35, 36
mineral insulated cable 71, 73, 108, 203
miniature circuit breakers (MCB) 56, 57,
  61
  advantages and disadvantages 63
  breaking capacities 63
  combined tripping 62
  and earth fault loop impedance 146
  magnetic tripping 61–2
  thermal tripping 61
  user misuse 198
Minor Works Certificate 128, 152, 153,
  159–60
  Guidance Notes 160
motion detectors 117
motor construction 244–5
motor control circuits 205
motor speed control 246–8, 258–9
  armature resistance control 258
  field control 258
  pulse width modulation
    (PWM) 258–9

motor starters 251–7
motors
  a.c. motor 233–43
  controllers 215
  faults 198
  on the job answer 316
  noise 259
  remote stop/start control 254
  slip 246, 247–8
  slip-ring motor 237, 257
  speed 246–8
  synchronous speed 235, 246, 247–8
  windings 246, 250
  see also a.c. motor; d.c. motor
moulded case circuit breakers
  (MCCB) 205
multicore cables 79
multimeter 35, 36, 304
multiplexer 117
mutual detrimental influence 165
mutual inductance 26

N

Neozed fuses 60
neutral conductor 66
NICEIC 152
no-volt protection 166
non-pressure water heater 111

O

off-load devices 192
ohmmeters 35, 36, 133, 150
  low-resistance 149, 184
Ohm's Law 36, 37
oil dashpot 62
oil-filled radiator 112
on-load device 192
open area (anti-panic) lighting 100, 101
operating devices 191
opto-coupler 290
out of reach protection 164
oven thermostats 114–15
overcurrent 196, 206
overcurrent protection 48, 57–68
  discrimination 68
  and earth-fault loop
    impedance 144–6
  fuses 58–60
  fusing factor 63, 64, 65
  operation of devices 195–6
  overload and short circuit 58
  Regulations 62–3, 64
  testing devices 157
  time/current characteristics 80, 81
overload current 196

overload and short circuit protection 58, 196

P

p-n junction 285–6, 292
passive infrared (PIR) sensors 117, 118, 198, 290
PC-based surveillance systems 117
perimeter protection 118
Periodic Inspection Report 152–3, 180–1
periodic inspection and testing 171–83
    detailed requirements 175–8
    frequency of 172–3
    general requirements 171–2
    on the job answer 315
    reasons for 171
    sequence of tests 178
    statutory and non-statutory documentation 173
    summary 183
    see also inspection and testing
permanent split capacitor (PSC) motor 242
phase angle 5, 18, 21
phasor diagrams 5, 6, 8, 9, 11–12, 14
    capacitance and resistance 20
    RLC circuit 16
photo cell 289
photodiode 289, 291
pillar terminals 202
plant failure 198
polarity 163, 184
    capacitors 280
    d.c. motor 230, 231
    testing 140, 157, 160, 178, 179, 183, 213, 215
pole pair 4, 230, 235
portable equipment: faults 208
potentiometer 271
power
    in a.c. circuit 19–25
    apparent 7, 10, 21, 22, 41
    measuring 39–42
    reactive 10, 20
    standby 109
    true 7, 10, 21, 22, 41
power factor 5, 7–8, 21, 22, 41
pre-commissioning tests 127, 131
programmers 98–9
Prospective Short Circuit Current (PSSC) 56, 66–7
protective conductors: continuity 178, 179, 213

protective devices 193, 194, 206–7
    inspection 124, 145–6, 176
    ratings of 195
Protective Multiple Earthing (PME) 49, 50–1
proximity switch 118
PVC covered MICV 203
PVC/PVC twin and cpc cable 203
PVC/steel wire armoured cable 203
Pythagoras' Theorem 22

Q

quad processor 117

R

radiant heaters 112
radio frequency interference (RFI) 259
reactive power 10, 20
reactive voltamperes (VAr) 20
rectification 285, 293–5, 306
    full-wave 293
    full-wave bridge rectifier 294
    half-wave 293
    smoothing 294–5
    three-phase rectifier circuits 295
Regulations
    cable selection 70–1, 72, 74, 75, 79, 80
    circuit breakers 62–3, 64
    electric shock 52–3
    emergency lighting 101, 102, 173
    inspection and testing 122, 123–4, 125, 130, 171
    lighting circuits 89
    residual current devices 55, 57
remanence 28
remote stop/start control (motors) 254
remote switching 205
repair process 217–21
    availability of replacement 217
    availability of staff 218–19
    and customer's premises 219–20
    downtime 218–19
    legal responsibility 219
    replacement cost considerations 217
Residual Current Circuit Breaker with Overload Protection (RCBO) 56, 158
residual current devices (RCD) 47–8, 54–6, 57
    labelling 177–8
    required sensitivity 55
    test instruments 151
    testing 55, 147–8, 158, 178, 180, 183, 215
    unsatisfactory test results 185

resistance 10, 19, 20
    a.c. circuits 5–6, 160
    measuring 35, 36
resistance, inductance and capacitance in parallel 18–19
resistance, inductance and capacitance in series (RLC) 16–17
Resistance/Capacitance (RC) circuit 14–15
Resistance/Inductance (RL) circuit 11–12
resistive circuits 5–6, 10, 19–21
resistors 11, 16, 19, 262–73
    coding 265, 266–7
    as current limiters 268–9
    fixed 262
    flashover 272–3
    light-dependent 273
    markings 265
    power ratings 271–3
    preferred values 264
    testing 268
    thermistors 273
    variable 263, 271
    for voltage control 269–71
rewireable fuses 58–9
ring circuit continuity testing 178, 185, 186
ring final circuit conductors continuity testing 213, 214
RLC circuit 16–17
rotor-resistance starter 257
rotors 232, 236–8
    squirrel cage 236–7
    wound 237–8

S

safety isolating transformer 31
Schedule of Test Results 152, 153, 155, 157–8
Screwits 201
Selection and Erection of Equipment 200
self-inductance 26
semi-resonant starter 92
semiconductor devices 285–92
    diode testing 292
    doping 285
    fibre optic link 290–1
    forward and reverse bias 286
    infrared source and sensor 290
    light emitting diodes 288
    opto-coupler 290
    p-n junction 285–6, 292
    photo cell and light-dependent resistor 289

photodiode 289, 291
zener diode 286–8
Separate Extra Low Voltage (SELV) 31, 156, 165
series motor 231, 258
series-wound (universal) motor 233–4
shaded pole motors 243, 249
shell-type transformers 29
short circuits 58, 66–7, 194, 196, 206
shower, electric 111
shunt motor 231, 258
silicon controlled rectifier (SCR) 306
simmerstat 113–14
sine wave diagram 2–3, 5, 8, 20
single-phase a.c. synchronous motors 249
single-phase induction motor 238, 245
slip-ring motor 237, 257
smoke detectors 107
socket outlets inspection 168–9, 175
soft starters 256
soldered joints 201
solid state equipment 206
space detection 118
space heating 112–13
    convection heaters 113
    direct acting heaters 112
    thermal storage devices 113
split-phase motor 239–40
squirrel cage rotor 236–7, 244, 245
standby power supplies 109
standby-lighting 100
starters: discharge lighting 91–2
storage batteries hazard 224
street lighting 200, 289
strobe light 118
stroboscopic effect 93–4
supply 182
    complete loss of 195
    localised loss of 195
surveillance software 117
switches 175
    faults 198, 204
    transistors as 301–3
switchgear 166, 204–5, 215
synchronous a.c. induction motors 249
synchronous speed 235, 247–8

T

terminating devices 201, 202
test instruments 148, 190
    test lamp 190
    test meter 190
testing
    non-live tests 213

thyristors 308
transistors 304
see also inspection and testing
thermal insulation 71, 72
thermal starter circuit 94
thermistors 273, 292
thermocouples 292
thermostats 110, 113–15
three-phase a.c. synchronous motors 249
three-phase induction motors 234–8, 244
    squirrel cage rotor 236–7, 255
    wound rotor 237–8
three-phase power 23–5, 39–40
three-phase rectifier circuits (d.c) 295
thyristors 306–10
    diac 309
    principle of operation 306–7
    testing 308
    triac 308, 309
timers 98–9
TN-C earthing system 51
TN-C-S earthing system 49–50
TN-S earthing system 48–9
transformers 26–34, 206, 293
    autotransformer 29
    calculations 31–4
    core-type 27–8
    current 30
    hysteresis loop diagram 28
    instrument transformers 29
    losses 27–8
    mutual inductance 26
    ratings 31
    safety isolating 31
    shell-type 29
    step-up and step-down 30, 34
    tapped winding 29
    voltage 30
    windings 27, 28, 29, 30
transient voltages 197, 206
transistors 296–304
    basics 296–7
    current amplification 299
    operation 297–8
    as switch 301–3
    testing 304
    voltage amplification 299–301
triac 308, 309
true power 7, 10, 21, 22, 41
TT earthing system 47–8
tubular heater 112
tungsten halogen lamps 87–8
two-pole voltage detector 190
Type 'D' fuses 60

U

ultrasonic intruder alarm systems 118
under voltage protection 166
under-floor heater 112
uninterruptible power supplies (UPS) 109, 224
unity power factor 7, 21

V

valence electrons 285
variable frequency drive (VFD) 259
variable speed drives (VSD) 259
video switcher 117
volt-amps (VA) 10
voltage drop 11
    cable selection 74–6
    and resistors 269–71
voltage indicating devices 190
voltage regulator diodes 286–8
voltmeter 21, 35, 36, 37

W

Ward-Leonard speed control 258
warning notices, inspection of 125
water heating 110–11
    earthing 110
    thermostat 110
wattmeter 21, 39, 40, 41
watts 10, 21
wiring accessories inspection 168
wound rotor 237–8

Z

zener diode 286–8
zenon light 118
zero phase angle 6

Buttercup's Pod

Fairy Godm

Fairy School of Magic & Music

SPARKLE TREE FOREST

Ice Cream Parlour

all the children in the world who
believe in fairies and magic. K.H.

...lary, who shakes the fairy dust. S.W.

First published in 2014 by Hodder Children's Books

Hodder Children's Books, 338 Euston Road, London, NW1 3BH
Hodder Children's Books Australia, Level 17/207 Kent Street, Sydney, NSW 2000

The right of Katharine Holabird to be identified as the author and Sarah Warburton as the illustrator of this Work
has been asserted by them in accordance with the Copyright, Designs and Patents Act 1988.

A catalogue record of this book is available from the British Library.

ISBN: 978 1 444 91338 5

Hodder Children's Books is a division of Hachette Children's Books
An Hachette UK Company
www.hachette.co.uk

Twinkle

Katharine Holabird and Sarah Warburton

Hodder
Children's
Books

A division of Hachette Children's Books

Twinkle's wings were
glowing with excitement!

"Tra-la-la-la-la!" she sang.
"I'm a little fairy
with sparkling wings.
When I get my magic wand,
I'll do amazing things!"

It was Twinkle's first day at
The Fairy School of Magic and Music
and she couldn't wait to learn fancy
spells like all the big fairies.

Twinkle skipped in line with all the little fairies waiting for wands.
Her wish came true when she got a sparkling peachy pink one!
Shiny fairy notebooks and glittery pens appeared on all the desks.

"We'll start with a simple spell," said Miss Flutterbee.

"Abracadabra, skiddle-dee-dee, this stool will now become a . . .

...tree!"

Before she could blink, Twinkle found herself perched way
up in a pear tree with Pippa and Lulu waving below.

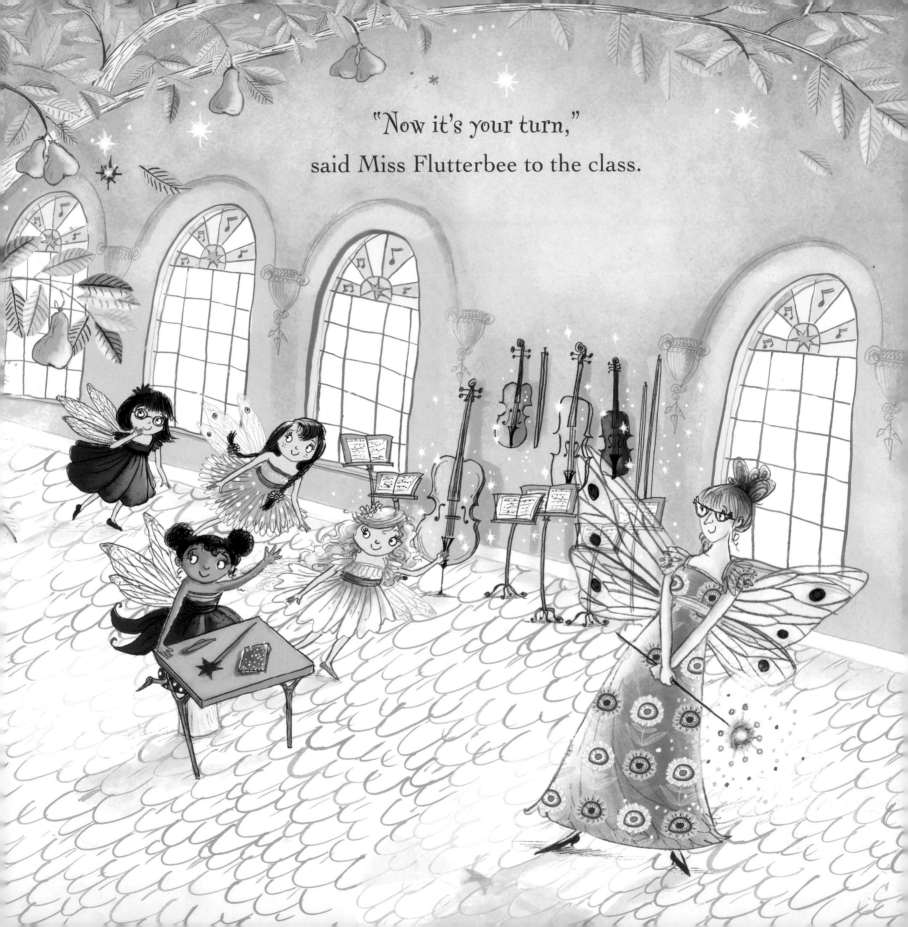

"Now it's your turn,"
said Miss Flutterbee to the class.

Pippa and Lulu swoooshed their wands and bingo!

They were both happily dangling from a plum tree.

Twinkle tried swooshing her wand but she got it all wrong, and instead of sitting in a plum tree she landed upside down in a prickly holly bush!

"Ouch!" cried Twinkle,
and Miss Flutterbee had to rescue her.

"Now remember to practise at home," said
Miss Flutterbee. "But **no** spells at night
– you don't want to start a ruckus."

That afternoon, Twinkle and her friends did their homework together.

"This is fun!" said Pippa, turning a teapot into a tiara.

"I love doing spells," said Lulu, changing bananas into balloons.

But Twinkle's swooshes were rather clumsy, and so her spells just got sillier and **sillier.**

"Don't worry, Twinks," said Pippa and Lulu, "we can practise again tomorrow."

Her wings fluttered wildly
as the bed began swinging,

the rocking chair
started rocking,

the alarm clock kept ringing,
and the lights were flashing.

Twinkle's cosy little fairy pod was shaking
and swaying, banging and clanging.

"HELP!" shouted Twinkle as the rumpus continued.

One by one all the creatures in Sparkle Tree Forest woke up, and they were not very happy… Bumble bees and ladybirds buzzed, grumpy gremlins and bullfrogs croaked, and groggy birds and bats fell out of the trees.

"Whooooo's making that racket?" screeched the owl.

"Stop the ruckus!" shouted the squirrel.

"Fiddlesticks and Fairy cakes!"
said Twinkle, waving her
wand faster and faster, but
the crashing and banging
only got louder.

The next morning, every creature in the forest was
howling and complaining outside Twinkle's door.

"Sorry about my silly spells," said Twinkle,
and she dashed off to see Miss Flutterbee.

of Magic & Music

"Gracious gremlins!" said Miss Flutterbee in surprise when she saw the parade of insects and animals following Twinkle.

"I'm having a spot of trouble with my homework," said Twinkle. "I keep forgetting the magic words!"

"Since you love to sing, try **singing** your spells instead.
Maybe that will help you remember," suggested Miss Flutterbee kindly.

"Brilliant!" said Twinkle,
and her wings glowed
bright pink again.

After school that day, Twinkle flew home singing the magic words.

"Abracadabra, skidddle-dee-dee, fix the mess that's up in my tree."

And when she arrived, Twinkle was overjoyed to
see everything was back in its place again.

As the sun set over Sparkle Tree Forest, Twinkle sprinkled fairy
dust over the treetops and sang a lullaby to all the forest creatures,
just to make sure they all had a really good sleep.

"Good night, sleep tight, with sweet dreams in your head.

Tomorrow we'll have lots of fun, but now it's time for bed!"

Lulu's Pod

Izzybell's Pod

Twinkle's Pod

Petal's Pod

Pippa's Pod

Gallery

Wand Shop

Café

Miss Flutterbee's Pod

Dress Shops

Library